D0123895

University of Alberta

INSECT
PESTS
OF THE PRAIRIES

by Hugh Philip
and
Ernest Mengersen

Alberta
AGRICULTURE

Alberta
ENVIRONMENTAL CENTRE

ISBN: 0-88864-870-7

THE PRODUCTION TEAM

Managing Editor:	Thom Shaw
Production Coordinator:	Val Smyth
Graphic Artist:	Melanie Eastley
Data Entry:	Arlene Connolly
Technical Reviewer:	Buck Godwin
Printing:	U of A Printing Services

Preface

This book is an expanded revision of "Insect Pests of Alberta" (Alberta Agriculture Agdex 612-1) published in 1975. The intent of the book remains the same -- to provide information on the appearance, life history, food hosts, damage and non-chemical control of the major nuisance and pest insects. The color photographs were selected on the basis of their value in aiding the reader to recognize a particular pest and/or the characteristic damage caused by a pest.

The text is divided into sections according to host categories (field crops, small fruits and vegetables, etc.), each of which contains descriptions of pests common to that host category. However, some pests are discussed as part of the description of a group of pests sharing similar characteristics (for example, aphids and scales), and are presented in one section only. For example, several scale insect pests of ornamental trees and shrubs are described in the greenhouse and houseplant section. Therefore, the reader is advised to consult the index of common and scientific names of pests or the host index to locate a particular pest in the text.

Hopefully this book will not only serve as a useful guide or reference for readers, but will also stimulate their interest and increase their knowledge and understanding of insects in general. The information presented in this book was collected from numerous extension and research publications. We are indebted to the many authors who have carefully studied insect pests and published the information which we have consulted and used freely.

Chemical control recommendations are not included because application methods and rates, use patterns, and products available are subject to change from year to year. Helpful suggestions are provided, where available, on non-chemical means of protecting yourself or your possessions from insect pests. Up-to-date information on the chemical control of insects and related pests is available from Federal and Provincial Departments of Agriculture and, in some municipalities, from local Parks and Recreation departments. Many bookstores and libraries have books offering additional information to that presented in this book.

Acknowledgements

Not only are we indebted to the many extension and research entomologists for the information they have collected and published over the years, but also to our colleagues who took the time to review and provide useful comments on this book during its preparation. Special thanks to B.J. (Buck) Godwin, Olds College for critically reviewing, commenting and discussing various aspects of the written work as it progressed, and to Dr. Bart Bolwyn of the Alberta Environmental Centre for editing the various drafts of the manuscript and for facilitating the publication of this book. Their encouragement and guidance helped bring this work to fruition.

We also wish to acknowledge the assistance from fellow entomologists Richard Butts, Alberta Environmental Centre; David Pledger, Alberta Environment; Dr. Garth Bracken, Agriculture Canada, Winnipeg; and staff at the Northern Forestry Centre, Canadian Forestry Service, Edmonton.

To Elsie Sakaluk and Elaine Cannan, our thanks for the many hours on the word processor.

We thank our colleagues across the prairies who so kindly supplied us with insect photographs for use in this book. A special acknowledgement to Marion Herbut for most of the photographic work published in this book. Specific credit is given under each photograph.

Finally, we thank our wives and families who were quietly patient and supportive during the long hours we neglected them while preparing this book.

We gratefully acknowledge all their help but the final responsibility for errors and omissions remains ours.

Hugh Philip

Ernest Mengersen

1989

Contents

Introduction

The number of known species of insects is approximately one million, well above the number of all other known species of animals. It is estimated that there are about 54,000 species of insects in Canada with a little more than half of these described and named. After 340 million years of constantly adapting to new and changing environmental conditions, insects have established themselves in about every terrestrial and freshwater habitat one can think of. Some species are found deep underground and some several hundreds of metres in the atmosphere. Others can survive several months exposed to below-freezing temperatures, whereas other species develop in water too hot to be borne by the hand. This extreme diversity in habitats is reflected by a similar diversity in the shapes, sizes, colors, behavior and food preferences of insects. For example, insects vary in size from 0.2 mm long to 160 mm long, and in shape from resembling a rose thorn to mimicking a leaf or twig or another insect.

Since recorded time, humans and insects have co-existed either because of overlapping habitats or because the activities of humans created suitable habitats for insect colonization. In their relation to humans, insects can be divided into two broad groups: beneficial and injurious.

Beneficial insects include pollinators that aid in fruit and seed production; producers of silk, honey, beeswax and a variety of other products; predaceous or parasitic species that assist in controlling some insect pest populations; species that are the sole or major food item for many birds, fish and other animals; species that aid in the control of noxious weeds; scavengers that devour or bury dead animals and plants. Many insects have an aesthetic or entertainment value as illustrated by some of our jewellery, paintings and tapestries. Scientists have used insects to help solve problems in heredity, evolution, physiology, ecology and environmental quality.

Injurious species include insects that damage or devour cultivated crops, stored grain and food, clothing, structures, and ornamental plants, and those that feed on humans and animals. Many pest species transmit internal parasites to humans and animals (e.g., house flies and dysentery), or may serve as primary hosts of diseases (e.g., mosquitoes and malaria). Some plant-infesting pests carry disease organisms (e.g., aphids and viruses) which can seriously weaken or kill plants.

A continuous battle is being fought with insect pests in an attempt to protect our homes, health, food and animals from insect ravages. The first attempts at protection were quite feeble and provided little relief. As technology advanced, chemicals were introduced (insecticides, acaricides) which provided some protection and reduced losses caused by insect pests. Initially these chemicals were derived from natural sources such as plants (e.g., nicotine, pyrethrum). However, synthetic chemicals were soon developed that enabled people to select from a broad range of products to apply against insects. Just as quickly as new chemicals were developed, insects developed resistance, rendering many products useless in situations where they were applied intensively. In recent years, public concern about the use of chemicals has intensified because of the persistence of some insecticides in the biosphere and the deleterious affects of chemicals on non-target organisms including humans. As a consequence, attention has been focused on developing safer and environmentally acceptable control techniques. These include the use of natural (biological) insect control agents such as predators and parasitoids (insects which parasitize other insects), viruses, bacteria and nematodes. The integration of natural control agents and chemicals has found widespread application in the control of certain pest species. Cultural control involves the modification and/or adoption of specific construction, cultivation, or management practices which will assist in reducing insect damage. Plant breeders are developing crop varieties that are tolerant or resistant to certain pest species, thus negating or reducing the need for chemical applications, and hence reducing costs of production. As a result of research into how insects grow and develop, scientists have identified and synthesized insect growth hormones. These hormones, when applied to immature insects (caterpillars, maggots, larvae), disrupt the normal growth pattern of the insects resulting in the death of the individuals. Discovery and development of new insect control techniques that are environmentally, socially and economically acceptable, will ensure continued protection of our possessions from insect depredation.

The remarkable ability of insects to invade and colonize various changing habitats around the world is better understood if one considers some of the attributes that have allowed insects to achieve their dominant position in the animal kingdom.

(1) SIZE: The relatively small size of insects has many advantages. Each insect requires little food thus allowing large populations to occupy small habitats where competition from other animals is minimal. They can find shelter from enemies and unsuitable environmental conditions in small cracks and crannies. Their small size also allows them to move about undetected; in some cases their presence is only indicated by the damage they inflict. A smaller body also makes possible great feats of strength and agility. For example, if a man had the same relative jumping ability as a flea, he could broad jump 215 metres and high jump 140 metres!

(2) REPRODUCTION RATE: Insects reproduce very quickly due to either
• production of a large number of eggs or young per generation, or
• a short life cycle and short interval between successive generations.
The time required to complete one life cycle in insects can vary from 10 days to

17 years. However, on the prairies, most insects complete one generation per year. Such factors as environmental conditions (temperature, humidity), predation, parasitism, disease, and availability of suitable food and egg-laying sites influence the reproductive potential of insect pest populations. The number of eggs a single female deposits in her lifetime can vary from one (e.g., true females of certain aphids) to millions (e.g., termite queens). Eggs can be laid singly over a long period of time, or in "batches" at certain intervals.

In most insects, reproduction depends upon mating between sexually mature males and females. The females subsequently lay eggs after an incubation period. Modifications to this bisexual system are present in many species of insects. In some, eggs complete development without even being fertilized (parthenogenetic reproduction). Eggs hatch in the female which then produces living young (e.g., summer aphids). Another example of parthenogenesis is displayed by honey bees and other hymenoptera. Unfertilized eggs give rise to males; fertilized eggs give rise to females. A further method of reproduction involves the production of two or more individuals from a single egg (polyembryony) which may or may not be fertilized (e.g., some parasitic wasps).

(3) METAMORPHOSIS: Newly hatched insects often do not resemble the adult form in either morphological characters or size. To acquire the adult form, immature insects must undergo changes in form (metamorphosis) which involves losing purely immature characters as well as acquiring distinctive adult features. All young insects undergo metamorphosis, whether this be simply growing (e.g., silverfish and springtails) or growing and acquiring adult structures. Three common types of metamorphosis are simple, incomplete (gradual), and complete metamorphosis.

Insects such as silverfish undergo simple metamorphosis. The young grow into adults with little change in appearance except for size and development of reproductive organs.

In the case of incomplete metamorphosis, the young closely resemble the adult except in size, color, shape, or presence of wings and genitalia. As the young (nymphs) mature, they gradually become more similar to the adult. Wings also grow, beginning as small wing buds or pads on the outside of the insect. As the nymphs grow, they undergo a series of molts during which they shed their rigid exoskeleton or "skin". The form of the insect between these molts is called an instar. Thus we can speak of certain species having four instars, seven instars, etc. With few exceptions, both adults and nymphs display similar habits, and live and feed on the same host. Insects undergoing incomplete metamorphosis include grasshoppers, aphids, plant bugs, bed bugs and lice.

Unlike the above-mentioned group of insects which require three developmental stages (egg, nymph, adult), insects undergoing complete metamorphosis require four developmental stages (egg, larva, pupa, adult). In this type of development, the young or larvae (e.g., maggots, caterpillars, grubs) do not resemble the adult. Wing buds develop inside the larvae in contrast to external wing development of nymphs. Larvae also require periodic shedding of the "skin" to grow before transforming into quiescent, inactive pupae.

Pupae are usually covered by a cocoon or some other protective coating. During this stage of development, extensive tissue reorganization occurs in which adult structures replace larval structures. It is during this stage that wing pads grow. Wings are not evident until the adult sheds the pupal skin and compresses its body to expand the wings to full size. Flies, beetles, butterflies, bees and wasps undergo this type of development.

Complete metamorphosis enables many insect species to colonize a wider variety of habitats thereby reducing competition for food and space as well as increasing the chances of survival and propagation.

(4) ADAPTABILITY: Insects continually evolve new structures and new habits or change old forms in response to constantly changing environmental conditions. This rapid accommodation to changing conditions is clearly illustrated by insignificant species becoming major pests through adaptation to new hosts (e.g., bertha armyworm attacking canola). The wide variety of habitats and hosts exploited by insects means that as a new situation is created at least one species of insects will undoubtedly be the first animal to occupy, adapt and colonize it. People are continually altering the physical and biological environment about them either by choice or neglect and thus are contributing to insect pest problems by providing new or improved habitats or by disrupting natural control processes.

(5) BODY STRUCTURE AND APPENDAGES: Another attribute of insects which has enabled them to achieve their dominant position in the animal kingdom is their unique body structure and appendages.

All insects are placed in the phylum Arthropoda which also includes spiders, lobsters, crayfish, crabs and shrimps. All these animals possess four common characteristics: segmented body, hard or leathery chitinous outer "skin" or exoskeleton, paired jointed appendages, and respiration by body surface, gills, air tubules (tracheae), or book lungs. The arthropods are further divided into a number of classes including the Class Insecta (insects), and Class Arachnida (ticks, mites, spiders).

The distinguishing characteristics of adult insects are:

(a) Three body sections (head, thorax, abdomen);
(b) Three pairs of jointed legs attached to the thorax;
(c) One pair of segmented antennae;
(d) Air breathing or respiration through air tubules (tracheae) and spiracles;
(e) Two compound eyes and often simple eyes (ocelli); and
(f) Two pairs, one pair, or no wings attached to the thorax.

Arachnids differ from insects in that they have four pairs of legs and two distinct body sections.

The insect head contains the feeding parts and sensory organs (antennae, compound eyes, ocelli). The thorax contains the locomotory appendages (wings and legs) and the abdomen contains the digestive and reproductive organs, and copulatory and ovipository (egg-laying) appendages.

Insect mouth parts are divided into two main groups – the chewing types and the sucking types. Insects with chewing mouthparts are able to bite off and chew their food (e.g., caterpillars, grasshoppers, beetles). Sucking mouthparts may be of several types: siphoning (e.g., butterflies, moths), rasping-sucking (e.g., thrips), piercing-sucking (e.g., mosquitoes, aphids, plant bugs), and sponging or lapping (e.g., house flies). Insects possessing sucking mouthparts feed on liquid foods such as plant sap, nectar, blood and digested tissue. Both sucking and chewing mouthparts consist of the same appendages which are reduced, modified, or lost depending on the food and the feeding habits of the insect. These appendages are the mandibles (jaws), maxillae and labium (lower lip). The hypopharynx, a tongue-like structure arising from the floor of the mouth-cavity, and the labrum or upper lip, are also associated with the mouthparts.

Insect antennae or "feelers" vary greatly in size, form and function, and are important characteristics in classification of insect species. The size and form of antennae is determined by the number and shape of the segments making up the appendages. Some are long and threadlike or short and feathery. Others resemble a string of beads, or a club, narrow at the base and gradually enlarging or abruptly knobbed at the end. Sometimes they are short with few segments and inconspicuous. Antennae are used by insects to feel their way, detect danger, locate food, find mates and communicate with others of their own kind. In general, the antennae are used to sense out the environment.

The compound eye is not a single eye but is composed of many circular or hexagonal areas called facets. Each facet is the lens of a single eye unit. The compound eye is especially good at detecting moving objects, however, it cannot move, focus, or perceive the shape of objects unless only a few feet away. The simple eyes or ocelli, located on the top of the adult insect head, are single eye units and distinguished by the arched and thickened transparent lens. These vary in number (maximum of three) according to species or may be lacking altogether, and serve to detect changes in light intensity.

Another reason for the great success of insects is their excellent mobility derived from their three pairs of legs and one or two pairs of wings. Anyone who has attempted to catch a grasshopper, cockroach, dragonfly, or house fly can appreciate their locomotory prowess. The thorax is divided into three segments – prothorax, mesothorax, and metathorax, each of which bears one pair of legs. Each leg consists of five independently moveable segments arranged in order from the base: coxa, trochanter, femur, tibia and tarsus. The tarsus or "foot" is composed of one to five segments or "knuckles". The terminal segment usually has a pair of claws and pads. In addition to walking and running, insects' legs are variously modified to serve such functions as digging, jumping, grasping, clasping, swimming, carrying loads and sound production.

Wings impart a high survival advantage to insects in that they enable insects to flee from danger and predators, forage far and wide in search of suitable food, disperse and colonize new habitats and often select nesting sites not accessible to predators.The wings are attached to the thorax – the front pair (forewings) to the mesothorax, the hind pair (hind wings) to the metathorax. Insects possessing only one pair of wings retain the forewings, with the hind wings either greatly reduced or missing. Insects which no longer require flight to survive have lost their wings (e.g., fleas). Wings vary a great deal in size, shape, texture and venation. Venation refers to the pattern of the wing veins or hollow ribs

which provide the structural framework of the wing. The classification of insects into orders is strongly influenced by the shape and texture of the wings. For example, insects possessing two pair of wings, the front pair thickened and leathery, or hard and sheath-like concealing the folded membranous hind wings are placed in the order Coleoptera (sheath-winged). Insects possessing only one pair of wings belong in the order Diptera (two-winged). Order Orthoptera (straight-winged) includes insects which have straight, narrow forewings (e.g., grasshoppers). Moths and butterflies have wings covered with fine, often brightly colored scales and are thus placed in order Lepidoptera (scale-winged).

The abdomen is generally the largest of the three body sections of adult insects and is typically 11-segmented, the eleventh segment often greatly reduced. Each segment sometimes bears a pair of small openings called spiracles through which the insect draws air to breathe. The oxygen-laden air passes through an intricate network of fine tubes called tracheae extending throughout the body of the insect. Carbon dioxide is removed via the tracheae. The abdomen also contains most of the circulatory and digestive systems, and all of the reproductive systems. In most insects, the abdomen terminates in external genitalia. These structures vary a great deal in size and form and are not always visible as they may be partly or completely withdrawn into the abdomen when not in use.

Unlike vertebrates such as humans in which the skeleton or body framework is inside the body, the skeleton of insects is for the most part outside the body and is referred to as an exoskeleton. The exoskeleton or "skin" is composed of three layers. The outer hard layer contains pigments and other substances including chitin which imparts a high degree of resistance to water, alcohol, dilute acids and alkalis. The body wall not only protects internal organs and tissues, but also bends inwards at various points to form supporting ridges and braces for muscle attachment. As previously mentioned, the insect body consists of a number of segments, and the body wall of each segment consists of a number of hardened plates or sclerites which are separated by sutures or membranous areas. Sutures are seam-like lines marking the infolding of the body wall which allow the hard body wall to bend. Membranous areas or lines permit the movement of various body parts and their appendages. Therefore, the insect body is not a simple, hollow, hard tube filled with life-support systems but rather a segmented tube made up of a number of light, flexible, strong, hard plates. The design and durability of the exoskeleton is one of the chief reasons insects have been able to live in the greatest variety of conditions.

(6) BEHAVIOR: The apparent fixity of purpose based on instinctive behavior is another important factor responsible for the success of class Insecta. Insects lack both reason and judgement. Everything they do and the way they do it is not learned but inherited. Their response to stimuli received from their surrounding environment is generally of two types: directed or complex. Directed responses consist of orientation of movement away from (negative) or toward (positive) the stimulus. The stimuli producing a directed response may be light, temperature, gravity, moisture, chemicals, touch or contact, currents of air or water. For example, light attracts many night-flying insects such as moths. Certain chemicals act as insect repellents by producing a negative response in some insect pests such as mosquitoes.

Complex behavior involves much more than a simple negative or positive response to a stimulus. Capturing prey, finding a mate, egg laying, building a nest, and making a cocoon are behavioral activities which require coordination of a number of movements and responses. These activities are not learned by individuals but are instinctive behavioral responses acquired hereditarily. Bees do not have to learn how to collect pollen, build hives, tend eggs and larvae, or protect the nest from intruders. Carrion beetles instinctively excavate beneath carcasses and bury them before laying their eggs. In many instances it has been necessary to learn the behavior of particular insect pests in order to design and implement effective control programs.

At first glance, the lack of ability to reason and form judgement would seem a disadvantage and place insects at our mercy. However, the wide variety of behavioral patterns of insects ensure that where one species is eliminated due to our efforts to exploit a particular behavioral response of that species, another species which behaves differently may eventually replace it. Individuals whose behavioral pattern is not beneficial to the species will gradually die out and be replaced by those whose behavior favors survival.

Therefore, the successful establishment and propagation of insects is a result of a number of morphological, physiological and behavioral attributes, some of which have been lightly touched upon in the preceding pages. Space does not permit a full discussion of the survival attributes of insects, nor is it the intent of this publication to do so. However, people interested in economic entomology should be aware of some of the reasons insects continue to challenge our attempts to improve and protect our food supply, health and possessions.

Insect Pests of Field Crops

Acari
Brown Wheat Mite
Petrobia latens (Müller)

Appearance and Life History

The brown wheat mite is a world-wide pest of cereals that has recently become a problem on the prairies. Adult mites resemble clover mites in size (about 0.5 mm in length), legs (yellow front pair elongated into long feelers) and color (brownish).

Brown wheat mites feeding on wheat

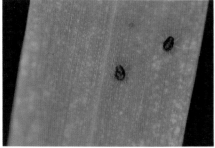

Lloyd Harris
Saskatchewan Department of Agriculture

The mites become active early in the spring and generally two or three generations are produced before mid-summer; consequently, populations can increase very rapidly. Only females occur; males have never been found. During spring and early summer, most of the eggs laid are nondiapausing reddish eggs that hatch in about 11 days. The six-legged larvae molt into eight-legged nymphs in 2-3 days. Two nymphal stages last about 2-3 days each and so adults are formed about 8-10 days after hatching. Adults live for 2-3 weeks. Towards the end of June, diapausing white eggs are laid that do not hatch for an indefinite period of time. Some will hatch in the fall if moisture is available, but most will hatch the following spring. Eggs, both diapausing and nondiapausing, are laid in the upper 50 mm of soil beneath soil clods, stones, or plant debris. Each adult mite can lay either about 80 nondiapausing eggs or 30 diapausing eggs. The mites thrive and multiply rapidly in dry weather and are often most damaging in dry years or after a succession of dry years.

Food Host and Damage

The brown wheat mite feeds primarily on monocotyledonous plants such as cereal grains and grasses but also feeds on various vegetables, melons, strawberries, legumes and fruits. On the prairies it is mainly considered a pest of wheat and barley. Damage appears to be similar to that caused by drought. Plants suffer from the loss of plant sap, which is withdrawn by the mites as food. Heavily infested plants appear to dry out even though sufficient moisture is available. A very fine mottling of the leaves and a bronzing or yellowing effect may be seen. This drying also results in a greatly reduced yield. The brown wheat mite is also a vector of a virus that causes a disease in barley called barley yellow streak mosaic.

This mite seems to be closely associated with crops grown on continuously cropped fields, especially barley continuously grown on saline soils.

Control

Control of the brown wheat mite may include crop rotation using nonsusceptible crops or the use of a pesticide early in the season when plants are still young and damage is minimal.

Coleoptera
Alfalfa Weevil
Hypera postica (Gyllenhal)

Appearance and Life History

The alfalfa weevil is an introduced pest of alfalfa. The adult is 4-5 mm long, from light to dark brown in color, and has a broad, dark brown stripe extending from the front of the head down the middle of the back for approximately two-thirds the length of the body. The head is extended into a long snout projecting downward, a characteristic of the beetle group known as weevils. Eggs vary in color from shiny yellow to brown and are inserted in clusters of 2-25 inside alfalfa stems.

Female alfalfa weevil laying eggs on alfalfa stem

Agriculture Canada, Ottawa

Newly hatched larvae are white and legless; mature larvae are 8 mm long, and green with a shiny black head and a distinct white stripe down the middle of the back. The tan-colored pupae are found in frail, oval lace-like cocoons spun among the leaves near the base of host plants.

Adults overwinter in protected sites often outside of fields, such as windbreaks, wooded areas, and under debris along fencelines. They may also survive under grass, trash in fields, or in the crowns of alfalfa. In the spring, after the alfalfa has started to grow, they take flight in search of food plants and egg-laying sites. Females chew holes in the stem before inserting the tips of their abdomens into the holes to lay clusters of eggs. Each female lays about 600-800 eggs during her life-time. Eggs hatch in 4-21 days, depending on the temperature. Larvae molt three times for a total of four larval instars. Larval development usually requires 3-4 weeks. Peak larval activity usually occurs from mid-June to mid-July although some larvae are present throughout the summer.

The larvae start to pupate in late June or July. The pupal stage generally lasts 1-2 weeks. Many larvae are killed during the first alfalfa cut but most pupae and adults survive. New adults begin to appear in early July, feed for a short period, disperse, and then remain inactive until September when they mate. In late October they leave the fields in search of protective sites for overwintering. There is one generation each year.

Food Hosts and Damage

Alfalfa is the main crop injured by alfalfa weevils but they may feed upon various clovers and vetches. Adults scar shoots by feeding and cause further damage by egg-laying punctures. These punctures are visible to the naked eye and indicate the level of egg-laying activity. Newly hatched larvae will feed for 3 or 4 days inside the stem before moving up the plant to feed on the developing leaf buds. Young larvae severely damage shoot tips by feeding within the folded leaves but the damage is not readily seen. Older larvae feed on the open leaves and skeletonize them, leaving only veins and stems. Defoliation is most severe toward the terminals, giving the entire field a greyish or frosted appearance. Notched leaves are characteristic of damage by weevil adults.

Damage is most severe to first-cut alfalfa but damage to second-cut alfalfa also occurs where the insect has been present for a number of years. When the first hay crop is cut, the larvae drop into the stubble and concentrate under the windrow for protection. While under the windrow, they feed on buds of the alfalfa crowns, retarding growth of the second crop.

The alfalfa weevil can reduce hay yield by 50 percent and the quality of hay is reduced. Feeding damage may also greatly reduce yield of seed fields.

Control

Field inspections to determine weevil population and stage of development are necessary for a successful control program. The number of weevils or egg punctures indicates the pest's abundance and expected damage. Early cutting of the first-growth alfalfa when the larvae are present will reduce populations of new adults, but this is not always practical. Green chopping, instead of cutting, windrowing, and baling can result in seasonal reductions of about 50 percent in total numbers of weevils. However, the stubble should be checked 3-5 days after cutting for signs of regrowth. If green-up does not occur, a stubble treatment with an insecticide should be made to control the surviving larvae.

A small wasp, *Bathyplectes curculionis* (Thomson), is important in biological control of the alfalfa weevil. In the past, this parasite was one of the main factors keeping the weevil under control. A predatory wasp, recently discovered in southern Alberta, may be beneficial in alfalfa seed fields. Proper soil fertility will also lessen the impact of larval feeding because of more vigorous stands.

Insecticides are recommended to control weevils when warranted.

Blister Beetles
Lytta spp. and *Epicauta* spp.

Appearance and Life History

Several species of blister beetles, often called caragana beetles, may be pests on the prairies. Blister beetles are easily recognized by their long, soft, cylindrical bodies, the flexible wing covers, and the rounded thorax which is narrower than the head and wing covers. The head is often pointed downwards. Blister beetles measure 12-25 mm in length. *Lytta*

species are quite conspicuous with their bright metallic bodies and iridescent wing covers. Eggs are yellow, elongate, and

Adult Nuttall's blister beetle on canola

H. Philip

more or less cylindrical. Larvae change considerably in form and appearance as they develop. When hatched they are active with relatively long legs that become shorter with each succeeding larval stage. Mature larvae are yellow, tough-skinned, about 13 mm long, and with mouthparts and legs greatly reduced. Pupae are white in color.

Adult beetles emerge from the soil in June and early July and congregate on food plants. Females leave every few days to lay eggs in the soil in a place likely to provide sources of food for the larvae. Each may lay four or five batches of 200-400 eggs. Eggs hatch in 2-3 weeks and the small active larvae feed mostly on grasshopper and cricket eggs or may attack the nests of ground-nesting leafcutter bees and bumblebees. The larvae then pass through four soft-bodied larval stages, feeding on egg pods or the pollen and honey stores of bees. Fully fed mature larvae molt to a hard-shelled larval form in which to overwinter. In the spring they molt again, reverting to larvae before pupating and finally emerging as adults. There is only one generation per year.

Food Hosts and Damage

The natural food plants of blister beetle adults are wild legumes, such as vetches, milkvetches and locoweeds. Among cultivated plants, fababeans and broad beans are highly favored foods. They readily feed on alfalfa, caragana, canola and coumarin-free varieties of sweet clover. Adults may move into the fields in swarms where they eat leaves and flowers, generally destroying the younger flowers and leaves before feeding on older parts of the plants. The amount of damage depends on the stage of growth of the plants at the time of attack and the size of the swarm of beetles. Populations of blister beetles often fluctuate with grasshopper populations and reach pest levels about 1 or 2 years after grasshopper outbreaks.

Bodies of blister beetles contain a toxic substance known as cantharidin which can cause painful blisters if a beetle is crushed on the skin. It has been reported that horses fed alfalfa hay with high numbers of blister beetles may suffer some digestive difficulties.

The larva of a brown blister beetle can reduce alfalfa leafcutter bee production by invading the leafcutter nest and destroying the pollen-honey provisions and the bee eggs.

Larvae of blister beetles are usually considered beneficial because many feed on grasshopper eggs.

Control

When blister beetles attack, control measures must be prompt and thorough. Insecticides give good control of these insects. Since seed set in some legumes is dependent on cross pollination by bees, the application of insecticides must be made when pollinators are not active.

Crucifer Flea Beetle
Phyllotreta cruciferae (Goeze)

Striped Flea Beetle
Phyllotreta striolata (Fabricius)

Appearance and Life History

The crucifer flea beetle and the striped flea beetle are common pests of cruciferous crops on the prairies. Crucifer flea beetles are 2-3 mm long black beetles with a bright bluish luster. This introduced species is the most common flea beetle in the grassland regions. Striped flea beetles are 2-3 mm long black beetles with two distinct wavy

yellow stripes along the back. This species, also believed to be introduced, is more common in the parkland areas across the northern edge of the prairies.

Adult striped and crucifer flea beetles feeding on canola seedling

M. Herbut

Flea beetle eggs are small (0.4 mm), elongate, and yellow. Larvae are a dirty-white color with brown heads and end plates. Mature larvae are slender, up to 6 mm long, with three pairs of short legs on the thorax. Pupae are white and 2.4 mm in length.

Beetles overwinter as adults under leaf litter along fence rows, shelterbelts and groves of trees. They emerge in April and early May and feed on volunteer canola and mustard, or on weeds such as wild mustard, flixweed or peppergrass. In the spring, flea beetles move to seedling canola or other cruciferous crops. Mating occurs in May and egg laying begins in late May and early June. Eggs are laid in batches of up to 25 on or in the soil, and the overwintered adults begin to die off in late June. The larvae feed on the roots of the developing canola for 3 or 4 weeks and are present from about mid-June to late July. After feeding, they form an earthen cell in which they pupate. Adults begin emerging in late July and early August and feed on the green tissue of suitable host plants that are still present. The feeding activity of these adults can continue into mid-October. However, by mid-September, most adults have entered a dormant, overwintering stage. Only one generation of flea beetles is produced each year.

Food Hosts and Damage

Canola and related cruciferous crops such as mustard, rutabaga, cabbage, turnips and radishes are the preferred hosts of crucifer and striped flea beetles. Most damage to canola is done by the overwintered adult beetles just before or soon after the seedlings have emerged from the soil. Flea beetles move into newly emerging canola crops from the borders by hopping when the temperature is below about 18°C and flying when the temperature is higher. They feed by chewing small holes in the cotyledons or leaves. Damaged plants typically have a "shot-hole" appearance when the tissues around the feeding sites in the cotyledons and leaves die. Losses from flea beetle feeding on young seedlings include reduced number of plants surviving, smaller and weaker plants, and delayed plant development. All contribute to reduced yield. That is especially true if the weather is hot and dry. Canola seedlings can withstand significant leaf area removal in the cotyledon stage under good plant growing conditions without significant reduction in yield. With heavy and continuous attacks, seedlings may wilt and die, particularly when feeding is combined with poor plant growth during hot, dry weather. Heavy infestations may destroy the entire crop and reseeding may be necessary. Once the crop reaches the three or four leaf stage, the plants are generally established and can outgrow the feeding damage. Also the number of adult flea beetles often begins to decline at that time.

Flea beetle damage to garden crops is similar in the spring but may be more serious in the fall because the leaves remain greener longer. Other species attack potatoes, peppers, tomatoes and sugarbeets but the life cycles and damage are similar.

Control

Very few cultural or preventive controls are available. Increasing seeding rates and allowing cruciferous weeds to grow in summerfallow fields may help in reducing damage if flea beetle populations are high.

The use of recommended seed treatments will provide some control and protection to the emerging canola seedlings. Daily inspection of newly emerged crops is necessary to identify flea beetle damage as it develops. If flea beetles are numerous on the plants or on the soil, if beetle feeding damage is over 50 percent of the cotyledon or leaf area, and if the weather is warm and dry, a foliar-applied insecticide may be necessary. If damage is only along the field margins, then controls may be restricted to damaged areas only.

Foliar-applied chemicals may be required to protect susceptible vegetable seedlings in the spring.

Prairie Grain Wireworm

Ctenicera aeripennis destructor (Brown)

Appearance and Life History

The prairie grain wireworm is native to the prairies. Adults, commonly called click beetles, are slender black beetles, 8-12 mm long, with backward-pointing projections near the middle of the tapered body. Click beetles are easily identified by the audible click they emit when placed on their backs. A hinged joint between the thorax and abdomen has a keel that allows them to arch themselves when on their back. By snapping down quickly they pop up into the air and flip over. Eggs are minute, oval and pearly white. Young larvae are white but change to a shiny yellow or tan to dark brown color as they grow. Larvae are slender, jointed, and usually hard-bodied. They have three pairs of legs behind the head and the last abdominal segment is flattened with a keyhole-shaped notch. Full-grown larvae reach a length of about 20 mm. Delicate white pupae are formed within earthen cells.

Adult beetles emerge in April and early May from the soil in which they overwintered. They become active when the air temperature is above 10°C, mate and then seek egg-laying sites. From late May through June, individual females will deposit 200-400 eggs in loose soil, or under lumps of soil. Depending on the moisture, temperature, and firmness of the soil, eggs are laid

Prairie grain wireworm larvae

H. Philip

anywhere from just below the soil surface to 15 cm deep. After 3-7 weeks, young wireworms hatch and commence feeding on the living roots or germinating seeds of cereals or grasses. If no food is found within 4 weeks of hatching, the larvae will die. Those larvae that survive their first winter can go for at least 2 years without any food other than humus. The larval or wireworm stage lasts anywhere from 4 to 11 years. They hibernate from 5 to 25 cm in the soil. Older larvae are commonly found feeding to a depth of 15 cm in the top soil. When full grown, usually in July, the larvae pupate 5-10 cm deep in the soil. Pupation lasts for less than a month, however adults do not emerge until the following spring. Larval activity is governed by temperature and moisture conditions. Cool wet weather forces the wireworms closer to the surface whereas dry hot weather forces them deeper into the soil. Also, cool weather restricts adult activity and lengthens the egg-laying period. Eggs laid near the soil surface or in compact soil are subject to high mortality when rapid fluctuations in moisture and temperature occur. Mortality is estimated at 92-98 percent in eggs and young wireworms. Most wireworm mortality occurs during the first 2 weeks of larval life. Mortality caused by bacterial diseases is especially high in soils of high moisture content.

Food Hosts and Damage

The prairie grain wireworm is considered the most destructive wireworm pest of grain on the prairies. It prefers grasses, both annual and perennial. It also attacks potato, sugarbeet, corn, lettuce, sunflower, canola and seed onions. The larvae feed on germinating seeds or young seedlings, shredding the stems

but seldom cutting them off. The central leaves die but outer plant leaves often remain green for some time. Damaged plants soon wilt and die, resulting in thin stands. Since thinning can also be caused by poor seed and dry conditions, many wireworm infestations are passed off as being poor seed or poor germination. Wireworms do most of their damage in the early spring when they are near the soil surface. During the summer months the larvae move deeper into the soil where it is cooler and where moisture is plentiful. Wireworms do not ingest solid plant material, but chew tissues, regurgitate fluids containing enzymes, and then imbibe the juices and their products that are made soluble by the enzymes.

Potato seed pieces are seldom damaged to the point where poor stands result. However, the new tubers can be damaged severely. Tunneling allows disease-causing organisms to enter and damaged tubers are less marketable.

Damage is generally higher in silt, medium textured, well-drained soils and in soils cultivated for at least 12 years. Damage is usually light in heavy or very light soils. Crops grown in newly broken sod can suffer great losses for 1 or 2 years; then the damage decreases rapidly only to gradually increase in succeeding years if no wireworm control measures are applied.

Whole potatoes buried in marked locations in a field in the spring or from early to mid-August will indicate whether or not wireworms are present. Bury the potatoes 10-15 cm deep for a couple of weeks, then dig them up and examine them for wireworm tunnels. Check the fields each year. Sieving several soil samples will also aid in determining the presence of wireworms.

Control

Treating seed with an insecticide for 2 consecutive years after breaking sod will normally reduce the problem to a level where further seed treatment will not be economically feasible. If summerfallow is part of the rotation, starve newly hatched wireworms by destroying all green growth during June and July. Work summerfallow as shallow as possible for weed control. Seed shallow and pack the seedbed to induce quick germination, and avoid very early or very late seeding.

Red Turnip Beetle

Entomoscelis americana Brown

Appearance and Life History

The red turnip beetle, a native North American insect, is an occasional pest of canola. Adult beetles are 7-10 mm long, bright red with black patches on their heads and pronotum and three distinct black lines running down their backs. Reddish-brown oval eggs, 1.5 mm in length, are found beneath clods of earth. Larvae are six-legged, rough-skinned, segmented black grubs that feed at night. The upper part of the body is a dark, smoky black, and the underside is brownish. When mature, the slow-moving, 12 mm long larvae enter the soil to form bright orange pupae.

Adult red turnip beetle

Agriculture Canada, Lethbridge

Red turnip beetles overwinter as eggs in the soil and hatch in late April and early May, shortly after the snow has melted and usually before canola is planted. The grubs feed on the foliage of volunteer canola seedlings and cruciferous weeds (mustard family) in May with some feeding occurring until mid-June. When full grown, grubs enter the soil and pupate at a depth of about 2.5 cm. The pupal stage lasts about 2 weeks. Adults begin emerging in early June and feed for 2-3 weeks, usually in the same fields in which the larvae fed. At the end of June, adults enter the soil for a one-month resting period. They reappear in late July and August and disperse to new canola fields. The beetles scatter throughout canola fields, and are often found mating near the tops of maturing plants. Females lay clusters of eggs near the base of food plants under leaf litter, lumps of soil and other objects, in shallow crevices, and in loose soil to depths of 6 mm. Adults may be seen until October. Only one generation is produced per year.

Food Hosts and Damage

Red turnip beetles are most commonly found in the aspen parkland region of the prairies and in the Peace River region of Alberta. The larvae feed on turnips, radish, cole crops, canola, yellow mustard, oriental and brown mustard, flixweed, shepherd's purse, wild mustard and several other cruciferous weeds. Adults have been found feeding on most of the same plants with a few exceptions. Neither larvae nor adults feed on stinkweed, the most common cruciferous weed in the fields. Potato, sweet clover, lettuce and beans have also been reported as hosts for adults.

Both larvae and adults feed on seedlings, leaves, stems, flowers and seed pods. They chew large ragged holes in the leaves. Beetles and larvae will feed on volunteer canola and cruciferous weeds in the area where eggs were laid the previous year until the food supply is exhausted or the field is cultivated. Generally, larvae are not a problem because they normally complete development before canola seed has germinated. Adults are responsible for the greatest damage. Newly emerged adults do not fly but migrate by walking to new fields in search of food. The beetles can move considerable distances to reach a canola or mustard crop. Feeding only on plants of the mustard family, they may move through a cereal crop and eat the cruciferous weeds and volunteer canola. Depending on the abundance of the insects, feeding damage will vary from the loss of small portions of the cotyledons and true leaves to complete defoliation and death of the plants. The beetles move slowly, completely devouring canola plants as they move toward the centre from the field margins, making the damage obvious from a distance. The march is usually concentrated in a moving front which is no more than a few metres wide. In August, the adults invade canola fields soon after emerging from their summer rest period and feed on the buds, flowers, pods, and stems of these plants. This feeding does not result in economic crop loss because the numbers of adults normally encountered in fields in August and September are not large enough to reduce the seed yield significantly.

Control

Crop rotation to a non-cruciferous crop is the first step in reducing losses caused by this insect. Fields with red turnip beetles should be cultivated in late fall or early spring to bury eggs and reduce larval survival. Early spring cultivation or spraying with an appropriate herbicide will remove cruciferous weed hosts and volunteer canola, thereby destroying the food plants of the larvae. The beetles have caused severe damage in canola fields situated next to fields where canola was underseeded the previous year to fescue or other forage crops and therefore not cultivated. Adult beetles can be controlled by spraying with a recommended insecticide as the insects enter a canola crop. Since they move in mass, one or two passes with the sprayer along the field margin, over and in front of the invading insects, will provide total control.

Sunflower Beetle

Zygogramma exclamationis (Fabricius)

Appearance and Life History

The sunflower beetle, a native pest of the sunflower, closely resembles the Colorado potato beetle. Adult beetles are 6-8 mm long and 4-5 mm wide. The head is reddish-brown and the pronotum (or shoulder) is pale yellow with a reddish-brown patch at the base. Each wing cover has three dark stripes that extend the length of the back. The fourth stripe ends at the middle of the wing in a small dot that resembles an exclamation point. That dot gave the sunflower beetle its scientific name. (Colorado potato beetles have five stripes on each wing cover.) The cigar-shaped eggs are yellow to orange in color and are about 2 mm long. The humpbacked larvae are dull yellow-green with a brown head and, when fully grown, are about 10 mm long. The yellowish pupae are about the size of the adult.

Sunflower beetle larva and adult

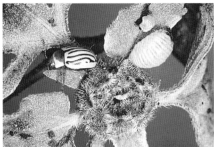

Agriculture Canada, Lethbridge

Sunflower beetles overwinter as adults in the soil and emerge in the spring about the time that sunflower seedlings begin to appear. After emergence, beetles feed on the emerging seedlings, mate within a week and begin egg laying. Eggs are laid singly on the underside of leaves or on the stem at a rate of about 15 eggs per day. Egg laying may continue for 6 or 7 weeks resulting in production of around 1000 eggs per female. Eggs hatch in about a week and the young larvae feed on the leaves at night and hide among the bracts of the flower bud and in the axils of the leaves during the day. Larvae feed for about 2 weeks but because of the long egg-laying period, larvae may be present in the field for about 6 weeks. When mature, larvae enter the soil and pupate in earthen cells. The pupal stage lasts about 2 weeks. Adults of the new generation emerge and feed for a short period in August and early September. They feed on the uppermost leaves or bracts of the plant before re-entering the soil to overwinter. There is only one generation per year on the prairies.

Food Hosts and Damage

Sunflower beetles feed on a variety of wild and cultivated sunflowers in the genus *Helianthus* across the prairies. Adults injure the plants by chewing on the edges of the newly formed leaves of the seedlings and may completely defoliate plants if the population is large enough. Larvae feed at night on the upper leaves and chew holes through the inner portion of the leaf rather than the margins as is done by the adults.

Feeding by adults and larvae may result in extensive defoliation, resulting in poor seed set or seed filling, reduced yields, and delayed maturity. Damage by the adults in late summer is not usually a problem.

Control

Natural controls such as ladybird beetles, lacewings, and other predators usually keep sunflower beetle populations below damaging levels. Insecticidal control may be necessary if defoliation by adults or larvae reaches 25-30 percent and if more defoliation will occur. One or two adults per seedling or 10-15 larvae per plant on the top 8-12 leaves will probably warrant control. Years when adult control is necessary will often also require larval control. One application of insecticide after all the eggs have hatched should give good larval control.

Sweetclover Weevil
Sitona cylindricollis Fåhraeus

Appearance and Life History

The adult sweetclover weevil is a small, 4-5 mm long, dark grey snout beetle or weevil found throughout the prairies. In the spring as the weather warms, weevils disperse by walking and flying. After mating in mid-May, females begin laying several hundred, white, sand-sized eggs on the soil near the base of second-year plants. Egg laying continues all summer. Eggs hatch in about 2 weeks. The small, greyish-white larvae burrow into the soil and attach themselves to the roots of the host plants where they complete their larval development. They live in the top 15 cm of the soil and will feed for 5-9 weeks before maturing. In late July mature larvae move up through the soil to pupate just below the surface. Adults emerge in mid-August and feed for a time before seeking overwintering sites under crop debris or in soil cracks. Thus, the weevils pass the winter in second-year sweet clover stands, and ditches and wastelands where volunteer sweet clover is commonly present.

Food Hosts and Damage

Sweet clover is the preferred host of this weevil, but it will feed on alfalfa or cicer milkvetch if no sweet clover is available. Adults chew crescent-shaped and jagged notches in leaves and can completely defoliate plants. They may even eat the outer tissue of the stem and green seeds in pods. Damage is most severe in dry years. Seedling crops can be severely thinned or completely destroyed if adults move into a field. Second-year stands can be thinned or stunted from the feeding activity of overwintering adults. Despite larval abundance and feeding on roots, plant growth does not seem to be retarded to any great extent. Sweetclover weevils drop from the plants very quickly if disturbed and are thus very difficult to spot.

Sweetclover weevil feeding damage to sweetclover

H. Philip

Control

Several management practices can reduce losses from sweetclover weevil. Arrange crop rotations so that successive plantings of clover are as far apart as possible. Weevils dispersing in spring and late summer will be less likely to find the first-year crop. Sow clover early (before grain crops) into a firm, moist seedbed and at the recommended shallow depth. That promotes even germination, a fast start, and hardy, vigorous seedlings. Cultivate clover silage and hay fields as soon as the crop is removed. Cultivation kills the larvae while they are still on the roots.

Inspect sweetclover seedings for weevil damage in the spring as the seedlings emerge. The weevils may not be seen, but the typical crescent-shaped feeding notches on the leaves are very noticeable. An application of insecticide over the entire field is recommended for protection of seedling stands if seedlings are being destroyed or severely damaged. In mid-summer and throughout August, inspect first-year clover stands for damage along crop margins. Invading weevils move into these stands only as far as necessary to satisfy their food requirements, so an insecticide treatment to affected field margins is usually all that is required.

Diptera
Hessian Fly
Mayetiola destructor (Say)

Appearance and Life History

The Hessian fly got its name from the Hessian troops that were used by the British during the American Revolution. The fly appeared shortly after their departure and it is suspected the fly was introduced into North America from Europe in the straw brought along for soldiers' bedding. It can now be found in most wheat growing areas of North America.

The adult Hessian fly is similar in appearance and about the same size as a mosquito (4 mm). The body is slender and generally black; females have a red-tinged abdomen. The wings are usually a dusky black. Eggs are small (1 mm long), slender, almost cylindrical and reddish in color. Newly hatched larvae are legless maggots, reddish-orange in color with a translucent greenish strip down the middle of the back. The larvae turn white as they feed and mature. The outer skin of the larva hardens and turns brown, forming a puparium, or protective case which resembles and is often called a "flaxseed". This brown, flattened "flaxseed" is very tough, allowing the insect to survive cold winters and hot dry summers.

There are probably two generations of Hessian fly each year on the prairies. The insect overwinters as a pupa attached to the base of leaves of some host plants. Flies emerge early in the spring as the mean temperatures reach 7-10°C. Emergence may take place over

Hessian fly puparium on winter wheat

M. Herbut

a 2- to 3-week period but the fly only lives about 3 days. On warm days the flies flit about, mate and lay strings of 2-15 eggs in grooves on the upper side of wheat leaves. A female may lay 250-450 eggs which hatch in 3-10 days. The maggots move to the base of the leaf sheath where they feed between the sheath and the stem for about 1 month. Once the larvae are full grown, feeding stops and the puparia are formed. A second generation of adults emerges in late summer and eggs are laid on young winter wheat or other host plants. The larvae feed and pupate before freeze-up in the fall.

Food Hosts and Damage

Wheat is the preferred host of the Hessian fly with winter and spring wheat both being susceptible. They may attack barley and rye but the damage on these crops is slight. Quackgrass and western wheatgrass are also hosts. Although the damage done to the alternate hosts is very small, they do serve to sustain the pest when wheat is not available. Injury to wheat is caused when the larvae feed on juices of the stem tissue at the crown or behind the leaves of young plants. The tissue around the point of attack becomes stunted and abnormal and may die. Stems injured in the fall usually die during the fall and winter, or are severely stunted. The leaves of infested plants are dark bluish-green and stand more erect than those of uninfested plants. Stems infested in the spring usually fall or lodge before harvest. Feeding injury may also cause shriveled heads and reduced yields. An examination of the leaf sheath and stem will determine if larvae or puparia are present.

Control

Hessian flies are minor and intermittent pests and control measures are rarely necessary. They require good moisture at specific times to undergo proper development so their survival in the dry prairies is often uncertain. The increase in irrigation and in winter wheat production may provide more favorable conditions for this insect. The switch to reduced tillage systems and continuous cropping may result in increased survival of Hessian fly. Prevention methods are easier than trying to control the insect once it is established. Plowing down volunteer wheat and stubble to bury the pupae and larvae is effective but may contribute to soil erosion. Planting winter wheat after about mid-September may prevent egg laying in newly seeded crops.

Orange Wheatblossom Midge
Sitodiplosis mosellana (Géhin)

Appearance and Life History

The orange wheatblossom midge is a recent pest on the prairies. The adult female is a fragile fly with a salmon pink body, about 3 mm long or one-third the size of a mosquito. The male is smaller. The head is light brown with two large black eyes. Legs are light brown and antennae are dark brown. Wings are dusky and fringed with hairs. The orange eggs are tiny and barely visible. Newly hatched larvae are white; mature larvae are oval-shaped and an orange-red color. Pupae are brown, oval-shaped, and found in the soil in cocoons smaller than canola seeds.

Orange wheatblossom midge larvae on wheat

Lloyd Harris
Saskatchewan Department of Agriculture

Adult midges emerge from the soil from mid-June to mid-July, about the time wheat heads are emerging from the sheath and beginning to flower. They mate and disperse through the fields. Females move downwind above the crop and then drop into the crop canopy and backtrack against the wind to find suitable egg-laying sites. During warm still evenings, eggs are laid on the wheat kernels singly or in groups of 3-5 under the edge of the glume or in the grooves of spikelets. Females lay an average of 30-40 eggs, with a maximum of 100 over their 4- or 5-day life span.

Eggs hatch in 5-7 days and the larvae move to the surface of the developing kernels and feed for 2-3 weeks. Mature larvae remain in the wheat head, each enclosed in a transparent skin, until activated by rain or damp weather conditions. Then they leave the head, drop to the soil surface and burrow into the soil to form overwintering cocoons. The larvae pass the winter in a resting stage in these cocoons. The following spring, if the soil is moist enough, they leave the cocoons, move about in the soil, and pupate near the surface. Adults emerge from the pupae about 2 weeks later to complete the cycle. If the soil is too dry, the larvae may remain in the cocoons to emerge in subsequent years when moisture conditions are more favorable.

Food Hosts and Damage

The major host is wheat; other grasses such as barley often have a low level of infestation. Winter wheat is also a host but it escapes serious damage if flowering occurs before the eggs are laid and if the kernels are fully developed before the larvae can cause damage. Eggs must be laid before the half-flowering stage for the larvae to survive and damage the kernels. Plants are most vulnerable to attack if eggs are laid during the time the heads are about one-half emerged from the boot to half-flowering (when the anthers are exposed). Larval feeding will cause the kernels to shrivel. Damage will not be evident unless the wheat kernels are separated and the developing kernels inspected. Three or four maggots per kernel will destroy that kernel. Damage is most severe in areas where high soil moisture in June allows the development of pupae and emergence of adults.

Control

Growing alternate crops such as canola, flax or field peas will prevent a population build-up of wheat midge. Since barley and winter wheat are not as severely damaged as spring wheat, they may be grown in areas where infestations are low. Parasitic wasps and mites may offer long-term control of this insect in dryer regions. Early seeding will help a crop develop beyond the susceptible stage before the flies emerge and begin egg laying. Insecticidal treatment against adults during the flight period may help reduce egg laying and subsequent damage.

Sugarbeet Root Maggot
Tetanops myopaeformis (Roder)

Appearance and Life History

The sugarbeet root maggot is a serious pest of sugarbeets. Adults are shiny black flies, about 6 mm long. Their transparent wings have a small dark area on the front margins. Eggs are white, slightly curved, slender, and laid in clusters of 10-20 in the soil around beet seedlings. The larvae are white, legless maggots that grow to a mature length of 12 mm. The oval, brown puparia are slightly smaller than mature maggots.

Adults emerge from the soil from late May to early July in fields that were planted to sugarbeets the previous year. They may be seen flying close to the ground on or near irrigation and farm equipment, flowering weeds and ditch banks. Mating occurs as male and female flies migrate to new beet fields. Egg laying begins about 10 days after emergence and eggs hatch in 1-3 days depending on soil temperature and moisture. The emerging maggots feed continuously on the developing sugarbeet roots. Mature larvae stop feeding in early August and remain dormant in the soil during the winter at a depth of 20-35 cm. The following spring they move to within 10 cm of the soil surface to pupate. One generation of sugarbeet root maggots is produced each year.

Food Host and Damage

Sugarbeets are the main host with lamb's quarters also being an important host. Young larvae feed on the roots of seedling sugarbeets, causing the seedlings to wilt and die when the tap root is severed. At thinning time, wilted plants are often the first noticeable sign of an infestation, especially on extremely warm days. The damage often

Sugarbeet plant killed by sugarbeet root maggots

Agriculture Canada, Lethbridge

results in reduced stands and, therefore, greatly reduced yields. Attacks on older plants can lead to secondary root growth and reduced size of harvestable roots. Heavy infestations, with up to 50 larvae per plant, can kill larger plants. Dark colored feeding channels, often accompanied by a rot, will develop on beet roots in July. Damage is most severe in late-seeded sugarbeet fields with losses as high as 50 percent in severe outbreaks. Crops planted in light to medium sandy soils are more prone to attack. Detection requires the digging up of suspected plants, and searching for white maggots either in the root or in the soil wherever root-tunnelling is observed.

Control

Cultural procedures can help reduce root maggot damage. Rotate crops so that beets do not follow beets where root maggots have been found. Growers are advised to plant early in a well prepared and well fertilized seed bed and to maintain adequate soil moisture through irrigation. Several granular insecticides are recommended for control when applied in the row during planting. There are presently no insecticides recommended that control the maggots once sugarbeets become infested.

Sunflower Midge
Contarinia schultzi Gagné

Appearance and Life History

The adult sunflower midge is a tiny tan-colored insect about 2 mm long with a wing-span of 4 mm. The wings are transparent without markings except for the veins. The larva is cream to yellowish in color and reaches a length of 3 mm at maturity. It is tapered at the front and rear and has no legs or head capsule.

The midge overwinters in the soil as a cocooned larva and pupates the following June. Adults emerge during the first half of July and live for about 3 days. Females lay eggs on the sunflower bud preferring those 25-50 mm in diameter. The yellowish-orange eggs are deposited on the bracts of the buds in clumps of 15 or more and can be seen on the plant without magnification. The emerging larvae migrate to the base of developing seeds or bracts and feed by rasping the plant tissues. As the larvae mature they make their way out of the head and drop to the ground for overwintering. There is usually only one generation per year; however, in some years, a partial second generation, of no economic significance, may occur.

Food Hosts and Damage

Although sunflower midge is a native insect it was unknown before 1971. Its occurrence is apparently restricted to commercial and wild sunflower and its distribution on commercial sunflower is confined to the Red River Valley where damaging populations have moved northward from North Dakota and Minnesota to the southern outskirts of the City of Winnipeg.

Both the position of the eggs on the bracts and of the bracts over the bud at the time of hatching have an important bearing on where the larvae locate within the head. Larvae hatching on bracts which are still closed over the center of the bud migrate downward to the base of developing seeds whereas larvae

hatching from partially or fully opened bracts migrate to the base of the bracts. Subsequent damage can range from complete destruction of all the seeds in the head to bract damage alone with no economic consequence. Destruction of seed causes the heads to distort as they grow giving them a cupped or clubbed appearance.

Sunflower head damaged by the sunflower midge

Agriculture Canada, Winnipeg

Control

Chemical control is not effective and insecticides cannot be recommended. Damage may be minimized by planting sunflowers during the last week of May or later. Planting adjacent to fields infested the previous year should be avoided.

Wheat Stem Maggot
Meromyza americana Fitch

Appearance and Life History

The wheat stem maggot is a minor but fairly regular pest of wheat on the prairies. Adults are yellowish-white flies about 3-4 mm long with three conspicuous black stripes on the thorax and abdomen and with bright green eyes. Larvae are slender, green spindle-shaped maggots, tapered at both ends, and measuring up to 7 mm long when full grown. The puparium is a slender pale green structure.

Wheat stem maggots overwinter as larvae hidden away inside the lower parts of the stem of wheat or other hosts. They pupate the following spring and adults emerge in June. After mating, the females deposit their eggs on leaf blades or stems of host plants on which the larvae feed. The young maggots crawl

Wheat stem maggot feeding inside wheat stem

M. Herbut

down inside the leaf sheaths to the tender soft part of the stems where they tunnel inside. When mature, the larvae pupate inside the stems. Adults emerge about mid-summer and lay their eggs on wild grasses or immature volunteer grain. The larvae of that summer generation become full grown by the end of August or during September. Two and possibly three generations are produced each year.

Food Hosts and Damage

The main hosts for wheat stem maggots are wheat, rye, barley and oats. They also feed on a number of forage grasses including timothy, bromegrass, crested wheatgrass and bluegrass. The head of an infested plant dies and turns white while the lower stem and leaves remain green. This condition is known as "whiteheads" or "silvertop". In the case of wheat, feeding takes place within the stem above the upper node for a distance of 5-8 cm. The injured stem is partly severed and the wheat head turns white and dies. The head and terminal straw of maggot-infested wheat will pull out easily due to the internal chewing damage by the larvae. Whiteheads are conspicuous in green fields but rarely exceed 1 or 2 percent of the heads in the field.

Control

Controls for wheat stem maggot are rarely necessary. Rotating wheat with nonsusceptible crops and the destruction of volunteer grains will reduce the pest population. Destruction of the infested straw by burning or cultivation will help reduce the numbers of insects. There are no effective chemical treatments for wheat stem maggot nor are there any resistant wheat varieties.

Heteroptera
Alfalfa Plant Bug
Adelphocoris lineolatus (Goeze)

Appearance and Life History

The alfalfa plant bug, an insect introduced from Europe, is an occasional pest of alfalfa grown for seed. Adults are 7-9 mm long, 2.5-3 mm wide. They may be distinguished from other plant bugs by their yellowish-green to green color, green legs and black spotting on the legs. The membrane on the front wing may be darker in color. They are about the same width as lygus bugs but are twice as long.

Adult alfalfa plant bug

Agriculture Canada, Lethbridge

Alfalfa plant bugs overwinter as eggs in the stem of their hosts. The eggs are slightly curved with a cap sticking out of the stem. They are clear when first laid but yellow as they mature. The nymphs hatch and feed during June and July. They are brown to light green and resemble the adults but have only wing pads rather than complete wings. The adults mate, feed, and lay eggs during the rest of the summer. There is generally only one generation per year, however, if egg laying occurs by mid-July, nymphs of a second generation may be evident in mid-August to early September. Migration of the adults seems to be a requirement for maximum egg production and movement from field to field or of several miles is common.

Food Hosts and Damage

Alfalfa plant bugs feed mainly on alfalfa, red clover and sweet clover. Several other herbaceous plants may be attacked but no native legumes are hosts. They are not economic pests of alfalfa grown for pasture, hay, or dehydration, because there is only one generation per year in western Canada and cultural practices prevent the development of heavy infestations in those crops. They attack alfalfa flower buds, causing the buds to shrivel, turn greyish-white and die. This is different from lygus bugs which normally attack later, feeding on flowers and young seeds. Alfalfa seed fields heavily infested with lygus bugs will blossom normally, but then will suddenly lose the flowers. An individual alfalfa plant bug will do more damage than a lygus bug.

Control

Controls are similar to those discussed under superb plant bug (Page 15). Because the adults fly long distances, an outbreak may occur suddenly. Regular sampling and sweeping is necessary to detect a sudden infestation.

Lygus Bugs
various species

Appearance and Life History

Lygus bugs are a complex of several species of bugs that can damage a wide range of plant hosts. Adult lygus are about 5 mm long and 2.5 mm wide. They vary in color from pale green to reddish-brown and have a distinct triangle or "V" mark about one-third of the distance down the back, just in front of the wings. Legs and antennae are relatively long. Eggs are slightly curved and approximately 1 mm long. Young nymphs are blue-green in color, and look like aphids but are more active. Color in the older nymphs becomes more variable and is similar to that of the adults. Nymphs develop prominent black dots on the top of the thorax and abdomen. There are five nymphal instars.

An adult lygus bug, *Lygus borealis*

M. Herbut

Lygus bugs overwinter as adults under debris, litter, or plant cover along fencelines, ditchbanks, hedgerows and wooded areas. In the spring, adults become active and feed on early-growing plants, mate, and migrate to crops when they become suitable for feeding and egg laying. This may be as early as mid-May on the southern prairies to mid-June in the Peace River region. Eggs are inserted individually into the hollow stems and leaf midribs or petioles of host plants. Egg laying usually lasts 3 weeks but may continue for up to 7 weeks or into early July. Eggs hatch between the end of May and mid-July when the crop is in bloom. Development from egg to adult takes about 45 days. Hot dry weather favors build-up of lygus populations and increases the possibility of damage to early growth. There are two generations per year on the southern prairies but only one in the northern areas. Adults remain in the fields until late summer when they move out of the fields to overwintering sites.

Food Hosts and Damage

Lygus bugs are general feeders found on many herbaceous plants. They are most damaging to alfalfa grown for seed but are also active feeders on canola. They also feed on such weeds as flixweed, kochia, lamb's quarters, mustard, Russian knapweed and Russian thistle. The tarnished plant bug, *Lygus lineolaris* (Palisot de Beauvois), feeds on a wide variety of forage crops, canola, vegetables, fruits and flowers, and is economically the most important lygus bug on the prairies.

Lygus bugs have piercing-sucking mouthparts and physically damage the plants by puncturing the tissue and sucking plant juices. The plants also react to the toxic saliva that the insects inject when they feed. Lygus bug infestations can cause alfalfa to have short stem internodes, excessive branching, and small, distorted leaves. Lygus bugs feed on buds and blossoms and cause them to drop. They also puncture seed pods and feed on the developing seeds causing them to turn brown and shrivel.

Canola damage is very similar as lygus adults actively feed on the base of the buds and flowers and cause blasting. Buds that are attacked appear shrunken and bleached white. When pods develop in late July and early August, the older nymphs and adults feed on the developing seeds by puncturing pods and sucking out the seeds' contents. Damaged seed appears dark brown and shriveled. A droplet of fluid may be seen on the exterior of the pod at the puncture site.

Control

Controls in alfalfa grown for hay are not necessary as natural predators and parasites, along with mowing and harvesting, reduce lygus bug populations. Chemical control may be necessary on alfalfa grown for seed. Where lygus has just one generation, or at most a small second generation each year, a single, well-timed application of insecticide to the alfalfa will provide protection for the entire growing season. Where lygus has two generations each year, two applications of insecticide may be necessary if weather conditions are conducive to the rapid development of an abundant second generation. Care must be taken to spray before leafcutter bees are placed in the field or to spray in the evenings with insecticides not harmful to the bees. Tarnished plant bugs can be controlled in ornamental plantings by removal of weeds and keeping lawns or grassy areas mowed to eliminate breeding sites.

Superb Plant Bug
Adelphocoris superbus (Uhler)

Appearance and Life History

The superb plant bug can be a problem on alfalfa grown for seed in southern Alberta and Saskatchewan. Adults are about 8 mm long, 2.5 mm wide, and bright red in color with a black stripe on the back and under surfaces. Legs and antennae are black. The antennae are about as long as the body. Eggs are cylindrical in shape, slightly enlarged in the midsection, with a cap at one end and rounded at the other. Eggs are pearly white when first laid but gradually turn pink and then red just before they hatch. The nymphs are usually dark red with less black than the adults.

Adult superb plant bug

Agriculture Canada, Lethbridge

Superb plant bugs overwinter as eggs in alfalfa stubble and straw left in the fields. Egg development does not begin until spring, and eggs hatch from the end of May to late June. During the month-and-a-half nymphal stage, the nymphs molt five times, each time growing in size and developing larger wing pads. Feeding continues through the nymphal and adult stages until fall when they die due to frost. Egg laying takes place during August and September on square, non-lignified portions of alfalfa stems with average widths of about 1 mm. The ends of older plants and thin second growth are preferred egg-laying sites. Eggs may be laid singly or in rows in the stem with only the cap sticking out. Populations normally build up slowly and are believed to take about 5 years to reach economic numbers in any one location. There is one generation per year on the prairies.

Food Hosts and Damage

Alfalfa is the principal host of this pest but it can also reproduce on Canada thistle. Both nymphs and adults feed by piercing plant tissue and sucking up plant juices. This results in reduced vegetative growth of the plant, destroyed or blasted flower buds, dropping of flowers, reduced pod formation, shrivelled seeds with poor germination and reduced seed yields. It is suspected that the saliva which is injected into the host tissue by the bugs while feeding may be toxic to the plant, killing the plant cells around the feeding punctures and resulting in reduced vegetative growth and destruction of buds and blossoms. Seeds that have been attacked are usually small, thin and greatly shriveled.

Control

Thorough burning of the alfalfa stubble and debris in late fall or early spring before alfalfa starts to grow destroys the eggs that were laid in alfalfa stems the previous summer and fall. Burning may control plant bugs in a field for more than 1 year. Normal cultural practices in haying or harvesting alfalfa for forage remove the eggs and prevent a build-up of superb plant bug populations. If burning of heavily infested fields is not possible, check the field with a sweep net to determine the size of the population and then apply a recommended insecticide when the alfalfa is in the early bud stage.

Homoptera
Grain Aphids
various species

Appearance and Life History

Aphids, or plant lice, are small, oval, fragile insects. Adults measure about 2 mm in length and are generally pear-shaped and pale to dark green in color. Antennae are usually quite long. Cornicles (two tubes projecting backward from each side of the abdomen) are characteristic of aphids. Aphids have sucking mouthparts. Adults may be winged or wingless with both forms present at the same time. Reproduction may be live births without mating in summer and by mating and egg laying in fall.

Aphids overwinter as eggs on a host plant. Wingless females or "stem mothers" hatch from the eggs in the spring and in about 3 weeks give birth to 50-100 live female young. These may remain wingless or become winged for dispersal. These second-generation females will begin giving birth to live female young in about 2 weeks. This process will be repeated every 20-30 days for the rest of the summer. As the days become shorter and the temperatures lower, males will be produced and mating and egg laying will occur.

Most species of grain aphids do not overwinter well on the prairies and so most economic aphid infestations originate from migrations on southerly winds from the southern U.S. each summer. Their reproductive potential is tremendous and large populations build up rapidly. Cool moist conditions seem to favor aphid populations, as food from plants is plentiful and predator and parasite activity is slowed, allowing aphids to reproduce unchecked.

The English grain aphid, *Sitobion avenae* (Fabricius), is bright green with black cornicles and antennae; legs are banded with areas of green and black; cornicles are long and narrow. It is found in colonies or scattered on leaves, stems, or in the heads of cereals and cultivated grasses.

Wheat head infested with English grain aphids

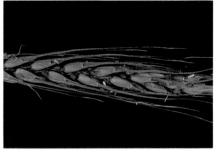

Agriculture Canada, Lethbridge

The corn leaf aphid, *Rhopalosiphum maidis* (Fitch), is greenish-blue with short broad cornicles with a dark spot surrounding their base; legs, cornicles, and cauda (a tail-like structure) are all black. It is found in colonies on stems and leaves of barley and corn, secreting much sticky honeydew.

The oat birdcherry aphid, *Rhopalosiphum padi* (Linnaeus), is dull olive-green with a reddish-orange patch between and at the base of the cornicles; legs and cornicles are pale green with black tips; antennae are black. It is found in colonies or scattered on lower leaves, stems, and roots of cereals and grasses.

The greenbug, *Schizaphis graminum* (Rondani), is pale to bright green with a dark stripe down the middle of the back; antennae are black; legs and cornicles are pale green with black tips. It is found in colonies on leaves of cereals. When feeding, a toxic saliva is injected into the plants causing discoloration and brown spotting at feeding sites.

Greenbugs feeding on wheat

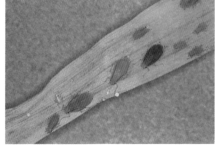

Agriculture Canada, Lethbridge

The Russian wheat aphid, *Diuraphis noxia* Mordvilko, is relatively small, lime-green in color, with an elongated, spindle-shaped body. It lacks the typical cornicles but possesses a projection above the caudal or tail segment which gives it a "forked tail" appearance when viewed from the side. Wheat, barley and triticale are preferred host plants. Damage symptoms include white, yellow or green longitudinal streaks along the leaves caused by toxins injected into the host while feeding. Other damage includes curled leaves and flattened, discolored plants that will eventually die if the aphids are not controlled.

Russian wheat aphids feeding on wheat

Lloyd Harris
Saskatchewan Department of Agriculture

Food Hosts and Damage

Grain aphids feed on cereals and grasses of various kinds. Direct damage to the crop is caused by sucking juices from the stems and leaves, removing valuable nutrients and causing fewer grain heads, lower numbers of kernels per head and reduced kernel weight. High populations of aphids on developing heads before the dough stage will remove moisture from the kernels to produce light and shrunken grain. Once the kernels pass the milk stage, aphids are unable to pierce the bran and damage no longer occurs. Aphid populations of 50-70 per head or plant before the soft dough stage may require controls. Seventy English grain aphids per head of wheat before the soft dough stage will reduce yields. High populations of corn leaf aphid on barley cause severe damage before the boot stage but not after. For Russian wheat aphids, economic damage can result with as few as 10 percent of young plants or tillers infested.

Aphids also carry and transmit barley yellow dwarf virus which can be important in barley and oat production. Barley yellow dwarf disease can reduce yields of dry forage and protein of oats and barley crops and reduce the height and number of tillers and as a result reduce grain yields. Late-seeded barley and oats grown for green feed or silage are the most susceptible crops for virus infection because of aphid population build-up and the presence of a susceptible crop. The oat birdcherry aphid is the main vector for barley yellow dwarf on the prairies.

Control

Natural predators such as ladybird beetles, hover fly larvae and lacewings usually keep populations of aphids under control. Early-seeded crops grow and become vigorous and pass the susceptible stage before aphid populations reach damaging levels.

Chemical controls are effective when carried out before the most susceptible crop stage is reached. Other factors to consider before planning a control program are (a) the species of aphid and its abundance, (b) the stage of development of the crop, (c) the presence of natural enemies, and (d) the potential crop value.

Pea Aphid

Acyrthosiphon pisum (Harris)

Appearance and Life History

The pea aphid is common wherever legumes are grown. Adult pea aphids are soft-bodied, slow-moving, and range in color from light to dark green. They are pear shaped, about 3 mm long and 1.5 mm wide, with long slim legs. Nymphs are smaller but closely resemble adults. Adults and nymphs have long slender cornicles.

Adult pea aphid feeding on alfalfa leaf

Agriculture Canada, Lethbridge

Pea aphids overwinter as eggs on leaves and stems of various perennial legumes. In spring, when the plants resume growth, a small, light green, wingless female hatches from each egg. These aphids, which are all females called "stem mothers", can reproduce without mating. They feed on the growing plants and give birth to female young. Some aphids of the second and third generations become winged and migrate to peas and other acceptable host plants. There they feed and produce wingless females that give rise to winged and wingless females.

Aphids develop from birth to maturity in 5-50 days depending on the weather. All pea aphids are female throughout spring and summer and a summer female can produce 50-150 young during her life. Whenever the host crop is cut during spring or summer, the winged aphids leave to search for new plants on which to live.

In late September or October, winged males and wingless females are produced. These mate and the females lay eggs. The eggs, which are deposited on leaves and stems, are yellow when first laid but soon turn green and then shiny black. Pea aphid eggs can survive low temperatures that kill other forms of the aphid. There may be 7-15 generations per year.

Food Hosts and Damage

Field peas, alfalfa and clovers are the main hosts of the pea aphid. Other legumes such as vetches, sweet peas, trefoil and beans are also attacked.

Pea aphids are found wherever peas and forage legumes are grown. Pea aphids mainly infest the growing tips of plants. Both the adults and the young suck juice from leaves, petioles, stems and flower buds. Infested pea plants become stunted, and yield as well as quality are reduced. Injury to alfalfa is distinctive and a heavily infested area in a field may be apparent from a considerable distance. The plants are stunted and wilted, the tip leaves are light green and the lower ones yellow or dead. From a distance the affected area appears brownish.The bare ground is readily seen and is usually covered with whitish molted skins of the aphids. Aphids are seldom a problem on the first cutting of hay but may reach outbreak numbers in late July or August before the second cutting. In alfalfa seed fields, aphid populations may build up in late August and early September but seldom need to be controlled as they usually do little damage at that time of year. Where alfalfa growth is retarded, weeds often take over and crowd out the alfalfa. High populations of aphids can also reduce cold hardiness of alfalfa. Pea aphids are also vectors of a number of viruses that cause plant diseases.

Control

Predators and parasites attack the pea aphid and help to keep it under control. Usually they become abundant only when the aphid is abundant. The predators are mainly damsel and minute pirate bugs, adults and larvae of ladybird beetles and lacewings, and larvae of hover flies. The parasites are the larvae of tiny wasps that develop in the aphids and kill them. When the aphids are abundant, spiders and birds also destroy them. A fungus disease may eliminate infestations of aphids in warm, moist weather.

Several pea varieties are not severely damaged by aphids. Some aphid-resistant varieties of alfalfa that have been produced in the U.S.A. will survive the prairie winter. Weather that is favorable for rapid growth of alfalfa greatly reduces the possibility of aphid damage. Aphid infestations may be reduced by very hot weather and retarded by cold weather. Heavy rains may dislodge and kill aphids.

Cutting alfalfa hay in the early bloom stage, removing the hay from the field quickly, and irrigating immediately reduces aphid damage to the plant regrowth because the crop becomes well established before the aphid population grows large.

Insecticides used for aphid control should be applied before the plants are severely damaged. Farmers should keep careful check on the build-up of aphid populations in their fields. One properly timed application of insecticide per season usually gives satisfactory control. Timing to prevent injury to pollinators in seed fields is very important.

Sugarbeet Root Aphid
Pemphigus populivenae Fitch

Appearance and Life History

The sugarbeet root aphid is an annual pest of sugarbeets. The life cycle is complex with at least two hosts and six forms. Winter is passed as eggs in cracks or under the bark of poplar or cottonwood trees. As the leaves begin to open in the spring, wingless females known as "stem mothers" hatch from the overwintered eggs. These aphids feed on emerging leaves causing them to bulge downward at the point of feeding along the midrib until a pocket or gall is formed around each aphid. Within the galls, the stem mothers give birth to 75-150 female young that are winged and have a black head and thorax and a green abdomen. Between late June and early August these winged forms fly or are blown in all directions. Those which land on sugarbeets give birth to many pale yellowish-white, wingless aphids that feed on beet roots. These summer forms are about 2.5 mm long and secrete a mass of white waxy material. The aphids are readily seen in the white material that is found on the infested beet roots and in the surrounding soil. Several generations are produced on the summer host each year.

Sugarbeet plant infested with sugarbeet root aphids

Agriculture Canada, Lethbridge

In late summer and early fall the wingless aphids on the beets give birth to large numbers of winged aphids which fly back to their winter hosts. These fall migrants are similar in form and color to the spring migrants. In protected places on the bark they give birth to several small, yellow, wingless aphids that are either males or females. After mating, the female deposits a single white egg in a crevice in the bark of a host tree or under the bark of a dead branch, where it remains for the winter.

It is only in fall that males are produced. Throughout the rest of the life cycle there are only females which give birth to live female young without mating, so populations can increase rapidly. Not all summer forms produce winged migrants in the fall. Some continue to produce wingless forms that overwinter in the soil in beet fields or on weeds in ditches and headlands and can become established on beets in spring.

Food Hosts and Damage

Overwintering eggs and stem mothers are found most commonly on narrowleaf cottonwood and balsam poplar, and occasionally on the plains cottonwood. The summer forms not only attack sugarbeets but also have been reported on red beets, swiss chard, spinach, alfalfa and lamb's quarters. Sugarbeet root aphids suck the sap from the roots of beets and can destroy most of the rootlets and severely damage the tap root. A few aphids will not appreciably change the appearance of the beets but large populations can cause the plant to wilt and, if enough sap is removed, the beets will die. As populations increase rapidly and are subterranean, fairly severe damage may be done before the aphids are noticed. Severe infestations not only cause the leaves to wilt and stunt the plants, but they also reduce the size, sugar content and quality of the beets.

Early frost damage is generally more severe on heavily infested beets. Infestations tend to be more severe if the soil becomes dry early in the season.

Control

The best method of reducing damage by the sugarbeet root aphid is to plant the sugarbeets early, irrigate them early and frequently, and keep soil fertility high to permit rapid, vigorous growth throughout the season. A number of aphid predators and parasites attack the sugarbeet root aphid in most stages of its life cycle.

Hymenoptera
Wheat Stem Sawfly
Cephus cinctus Norton

Appearance and Life History

The wheat stem sawfly, a native insect that lives in grasses in western North America, has often been a serious pest of wheat on the prairies. Adult wheat stem sawflies are shiny black, wasp-like insects about 8-13 mm long, with three yellow abdominal bands, smoke-colored wings and yellow legs. The adults can be recognized in the field by their habit of resting, head downward, clinging close to the stems of grain plants. The name "sawfly" is used for these insects because they use their sawlike ovipositor to cut a slit in the plant tissue to lay eggs. Eggs are crescent-shaped, glassy, milky white, and about 1 mm long. The pale yellow, newly hatched larvae are S-shaped with a wrinkled body. These worm-like larvae have a brown head and a short, blunt pointed projection at the tail end. Larvae may reach a length of up to 13 mm when full grown. The slender shape allows the larvae to live and bore within the stem of grasses. These characteristics along with the frass or sawdust that is found in association with the sawfly larvae serves to separate sawflies from other insects that infest wheat stems. Pupae are slender, about 12 mm long, and are found inside the bases of the stubble.

Wheat stem sawfly larva in overwintering position in base of wheat stem

H. Philip

Adult wheat stem sawflies emerge during mid-June to early July from stubble fields and native grasses. They are rather inactive insects that drift from plant to plant but spend most of their time resting on grass stems. They do not feed but do require water to mate and lay viable eggs. The adult female lays up to 50 eggs in immature wheat. There is usually 1 egg per stem, although several females may oviposit into the same stem. Eggs hatch in about 1 week. There is only one surviving larva per stem as the first larva to hatch eats the remaining eggs. Larvae feed on the pith of the stems and bore through the joints for about 2 months. In August, larvae migrate to the bottom of the stem, turn around with head upwards and cut a V-shaped groove entirely around the inside of the stem just above ground level. The exposed end is plugged with frass and the larvae prepare to overwinter in the stubble by building long thin brown transparent cocoons just below the cut. Pupation occurs the following May. Usually, there is one generation per year although a generation can last 2 years as very high springtime temperatures can force the larvae back into the resting stage.

Food Hosts and Damage

Wild grasses are the primary host plants of the wheat stem sawfly. Spring wheat and rye are the main cereals attacked. Eggs are laid in barley but rarely develop while oats and fall rye are immune. Some introduced forages also may serve as hosts. The greatest losses occur around the margins of the fields. Losses by the wheat stem sawfly are of two types. The first is caused by the larvae feeding within the stem of the plant. This reduces the yield of the crop and lowers the quality of the grain harvested. The second major loss is the stem-cutting of the larvae that causes the stems to break in the wind, fall to the ground and become unharvestable.

Control

A number of control approaches have been implemented to reduce the losses caused by this insect. The use of solid-stemmed, resistant varieties of wheat tends to keep down the severity of sawfly damage as the larvae have difficulty surviving in these wheats. Swathing sawfly-infested wheat as soon as kernel moisture drops below 40 percent is also recommended to save infested stems before they fall.

Alternate crops immune or resistant to the wheat stem sawfly such as barley, oats, winter wheat or non-cereals such as flax or alfalfa, may be grown instead of wheat. A temporary trap crop of a susceptible variety placed around the crop and then swathed for hay or silage in July may help to remove some of the larval population. A permanent trap crop of smooth bromegrass around a field will reduce the number of larvae surviving in the ditches and headlands. Delayed seeding in the spring produces a crop that is unattractive to females at egg-laying time. Late maturing varieties allow two generations of parasites, resulting in fewer sawflies the following year.

Summerfallowing infested stubble and cultivating in early June to bury the larvae has been beneficial. Shallow tillage in the fall will increase larval mortality and help reduce the sawfly population. Deep tillage will bury the overwintering larvae and reduce adult emergence dramatically but the possibility of soil erosion is too great a risk for that practice to be recommended. Burning of the stubble reduces sawfly larvae but also greatly reduces parasite numbers.

Continuous cropping of susceptible crops and reduced tillage may increase the possibility of larval survival and therefore increase infestations in the future.

Lepidoptera
Alfalfa Looper
Autographa californica (Speyer)

Appearance and Life History

The alfalfa looper is occasionally a serious pest of field and garden crops. Adult moths have silvery-grey forewings marked with a distinct yellow, sickle-shaped spot near the middle of each wing, while the body and hind wings are dull grey or brown. Moths have a wing-span of 30-38 mm. Eggs are round and white to cream-colored. Mature larvae are about 25 mm long, light green to olive-green in color with a paler head, and a light stripe down each side and two light stripes along the back. Mature

larvae appear to have a swollen abdomen. Larvae have three pairs of legs on the thorax and three pairs of prolegs on the abdomen (one pair on segments five and six and one pair on the terminal segment). Larvae move in a looping motion due to this reduced number of prolegs.

Alfalfa looper larva

M. Herbut

Alfalfa looper adults are blown in from the U.S. each year, although some overwinter as pupae in the soil. The moths appear all summer long due to two overlapping generations. They feed on flower nectar at dusk and fly during the daylight hours. Eggs are laid singly or in small groups on the undersides of leaves of host plants, usually near floral parts. Each female can lay 150-200 eggs. Larvae hatch in about a week. After 4 weeks of feeding, the mature larvae attach themselves to a plant and spin a woolly cocoon in which to pupate.

Food Hosts and Damage

Preferred food plants of the alfalfa looper are alfalfa, clover and lettuce but they will attack canola, various vegetables and ornamental and tree fruits. Damage to canola is not common but has occurred in northern and southern Alberta. Larval feeding damage is characterized by clipping of flowers and small seed pods, ragged holes in the leaves and on the leaf margins to total defoliation. Canola plants can recover from light damage. Few alfalfa loopers survive to the pupal stage since death usually occurs in the fifth to sixth instar due to a virus. Unfortunately, damage to canola may be done before viral control can be effective.

Control

No cultural controls for alfalfa looper presently exist. Alfalfa loopers can be controlled with a recommended foliar insecticidal spray. If an infestation occurs, assess the damage and delay spraying as long as possible to allow diseases an opportunity to control the pest. Avoid application of insecticides when pollinators are present.

Army Cutworm
Euxoa auxiliaris (Grote)

Appearance and Life History

The army cutworm is a sporadic pest of crops on the prairies. The adult moth is a large (40-45 mm) grey-brown "miller" with two prominent spots on the forewing. Larvae are pale greenish-grey to brown with the back pale-striped and finely mottled with white and brown but without prominent marks. There is usually a narrow pale mid-dorsal stripe. The skin

Army cutworm larva

H. Philip

texture consists of fine, close-set, irregular granules. The head is light brown (tan) with small dark brown spots. Full-grown larvae may reach 40 mm in length. The pupae are indistinguishable from other cutworm pupae.

The moths emerge over a short period in June and after a short activity period find shelter in buildings or brush where they escape the hot summer temperatures. In September the moths resume flight activity and mating and egg laying occurs. Eggs are laid in or on loose soil during late afternoons in September and early fall. Each female can lay 1000-3000 eggs, an important factor contributing to the outbreak potential of this species. The presence of vegetation

may or may not be a requirement for egg laying. Hatching occurs in the fall when moisture conditions are favorable. The larvae feed almost entirely on leaves and stems of host plants but spend non-feeding periods below the soil surface. Very little crop damage is noticed at this time. Larvae overwinter in the soil and resume feeding in April. It is at this time that most economic damage occurs. Their seasonal occurrence in April and early May is their most distinctive characteristic. Pupation occurs about 7 cm in the soil in an earthen cell and lasts about 4 weeks in late May and early June. Only one generation of army cutworms is produced each year.

Food Hosts and Damage

Army cutworms feed on a variety of plant hosts including cereals, alfalfa, native grasses, corn, sugarbeets, sunflowers, sweet clover, vegetables, weeds and most notably winter wheat and fall rye. Holes eaten through the leaf and semicircular notches eaten from the edges of the leaf are the first signs of damage. Feeding is done entirely above ground and after the leaves are removed from one plant, the larvae will move to adjacent plants. When their food supply is depleted, larvae will move in large masses to new areas, hence the name "army" cutworm. Migrations up to 5 km have been recorded. Larvae generally feed during the late afternoon and early evening. Outbreaks may appear very suddenly and are preceded by a year with a dry July and a wet fall. Since moisture is required for eggs to hatch, germination of winter annual crops and weeds and egg hatch occur at the same time, providing food for the new larvae. An abundance of moths in June does not necessarily mean a cutworm outbreak the following year. A dry fall, especially during September, will reduce the number of overwintering larvae by delaying egg hatching and by killing eggs and newly hatched larvae through desiccation. Outbreaks can range from single fields to thousands of hectares. Severe infestations are characterized by complete defoliation of the food plants.

Control

Since the moths prefer freshly worked soil, a crusted surface during the egg-laying period in early fall may help to protect individual fields. Delaying seeding in the spring until late May when the larvae have completed feeding and begun to pupate is often effective. However, this practice is ineffective during a cool spring that slows larval development. Forage crops and pastures must be closely watched in April and early May for the presence of these larvae. Plants which have adequate moisture and are vigorously growing with 12-15 cm of top growth can withstand four larvae per 30 cm of row without loss of yield. Insecticide treatment is recommended if two or more larvae per 30 cm of row are present and the plants are under 10 cm in height.

Armyworm
Pseudaletia unipuncta (Haworth)

Appearance and Life History

The armyworm gets its name from its marching habit when looking for food. Adults are uniformly light brown moths, about 25 mm long with a wing-span of 35-45 mm. There is a small but prominent white spot in the center of each forewing. Eggs are greenish-white; young larvae are pale green in color and have a looping crawl like inchworms until about half grown. Older larvae are greenish-brown with numerous longitudinal white, pale orange and dark brown stripes. Each proleg has a dark

Armyworm larva

Lloyd Harris
Saskatchewan Department of Agriculture

band on its outer side. The head is pale brown with a green tinge and mottled brown appearance. Full-grown larvae are about 40 mm in length. Pupae are brown and resemble other cutworm pupae.

Armyworms may overwinter as partly grown second-generation larvae sheltered in the soil around clumps of grass or under litter on the ground or may migrate as adults on wind currents from the southern states early in the year. In the spring, larvae resume feeding until they are full grown and then enter the pupal stage in late April. Adults emerge, mate and lay eggs in May and June. A female may lay 700-1400 eggs in rows or bands of a few to several hundred in folds of leaf blades and under sheaths of grasses in dry stubble or dead leaves. Hatching occurs in 1-3 weeks and the young larvae feed in groups at night, hiding by day under stones, lumps of soil or in the center leaves of plants. Feeding continues for about 4 weeks until larvae are full grown. Pupation takes place in the soil and may last about 2 weeks. There are usually two generations per summer in Canada with the first generation doing the greatest damage in late June and early July.

Food Hosts and Damage

Armyworms feed primarily on members of the grass family, damaging crops such as oats, wheat, corn, barley and forage grasses. Broadleafed plants such as alfalfa, cabbage and turnips may also be attacked but the damage is minor. Larvae feed on the leaves, stripping the leaf margins and move up the plants to feed on the panicles and flowers, stripping off the awns and kernels. Dense, lodged cereal crops are more susceptible to damage from armyworm larvae. Lodged plants provide a favorable habitat for moth concealment and egg laying, and also provide the high humidity required for establishment of the larvae. The best time to look for armyworm larvae is in the evening or early morning when they are feeding up on the plants.

Control

Controls are rarely necessary in Canada but severe local infestations may require action. Five or six armyworms per square foot (50-65 per square metre) throughout an area in a field would warrant chemical control. Only infested areas of the field should be treated. Spraying should be done in the evening when armyworms are feeding on the plants. There is no benefit in applying a chemical once the armyworm is nearly full grown, pupae are present, parasitism is extensive or the crop is nearing maturity. By that time most of the damage will have been done.

Beet Webworm
Loxostege sticticalis (Linnaeus)

Appearance and Life History

The beet webworm has been a sporadic, and at times, a serious pest on a wide range of field and garden crops on the prairies. Adult webworms are greyish-brown moths about 20 mm long with a wing-span of 25 mm. When at rest the moth appears triangular in shape with a characteristic white margin on the end of the wings. Eggs are small, creamy yellow, and disc-shaped. Newly hatched webworms are pale green caterpillars about 1-2 mm long. Mature larvae are 25-40 mm long, slender and quite active. They are olive-green in the early stages and become black with maturity. There are two white or cream-colored stripes on either side of the black centre line of the back plus two rows of paired circular marks down either side of the back. Pupation takes place in long, soil-covered tubular silken cocoons which the larvae construct in the soil.

Beet webworm larva

M. Herbut

Beet webworms overwinter either as pupae or larvae within cocoons. Moths first appear in late May or early June. They normally fly at night, however, if disturbed from favored patches of plants, they will readily make short, rapid flights in a zigzag pattern. Eggs are laid overlapping one another like shingles, on the underside of leaves on preferred host plants. After 7-10 days, larvae hatch and feed on the undersides of leaves. They spin silky webbing at the tips of plants and will hang by silk threads from the leaves if disturbed. Larvae will often migrate in "armies" to nearby crops when weed hosts are destroyed by defoliation, drought or herbicides. Four to 5 weeks are required before they become full grown and drop to the ground to construct silken cocoons.

There are normally two generations each summer. The first generation of moths generally appears between mid-May and early July and the second generation of moths appears from mid-July to the end of September. There is some overlap between the two generations. The most serious damage to crops is caused by larvae of the first generation. Second-generation larvae feed and pupate before winter. Some overwinter as larvae and pupate the following spring.

Food Hosts and Damage

Beet webworms have been found on a wide range of broadleafed crops and weeds. The preferred plants include lamb's quarters, Russian thistle and sugarbeets. As pests, they are most damaging to canola, flax and sugarbeets. They have also been reported on sweet clover, alfalfa, mustard, sunflower and various vegetables. The larvae start feeding on the leaves of canola, and then move to the stems and pods, stripping surface tissue and giving the crop a whitish appearance, usually in local areas within the field. Damage often results from invasion of beet webworm larvae that have developed in an adjacent weedy field. Such an invasion may completely destroy the invaded portion of a crop. Crops with light infestations may suffer reduced yields from pod peeling.

Outbreaks generally occur in years of hot, dry summer weather when the weed hosts dry up causing the webworms to migrate into nearby crops. Hot weather also increases their food intake and contributes to a rapid rise in population. The affected crops not only suffer from drought but also rapid defoliation. A related pest, the alfalfa webworm, *Loxostege cerealis* (Zeller), attacks canola causing similar damage as the beet webworm. The larvae closely resemble beet webworm larvae in appearance and size.

Control

Beet webworm larvae are attacked by a number of parasitic insects which can reduce the numbers of moths of the second brood to below an economic threshold. Pupae in sandy soil are more susceptible to parasitism than pupae in clay soils. Crops, especially sugarbeets, should be checked carefully for eggs and young if moths are readily observed in the field. Large moth flights are not always followed by high numbers of webworms. When counts of webworms reach 1 or 2 per sugarbeet leaf on over half the leaves, control measures should be taken. Economic thresholds on canola are similar to those for bertha armyworm. When a field is heavily infested the use of insecticides is the only satisfactory control method. For effective control, insecticides should be applied before the webworms are half grown. Control of weed hosts helps to limit webworm populations in crops.

Bertha Armyworm
Mamestra configurata Walker

Appearance and Life History

The bertha armyworm is an important pest of canola on the prairies. Moths are predominantly dark colored with a patch of pale green on the thorax and at the base of each wing. A silvery spot and silver fringe on the forewings are also characteristic. They have a wing-span of about 38 mm. Pinhead-sized eggs are white when laid, and then gradually darken as the larvae develop inside the eggs. Newly emerged larvae are small, 3 mm long, pale green with a pale yellowish stripe along each side. The larvae molt

six times and pass through color phases of green and pale brown before becoming large velvety black caterpillars. Mature larvae are 4-5 cm long, with a light brown head and a broad pale orange stripe along each side. However, some remain green or pale brown throughout their larval life. Pupae are reddish-brown and resemble other cutworm pupae.

Bertha armyworm larvae

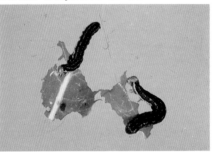

M. Herbut

Bertha armyworms overwinter as pupae in the soil. Moths emerge from overwintering pupae from mid-June to mid-July. Soon after emergence, mated females lay eggs on the undersides of canola leaves in clusters of 50-150 in a honeycomb arrangement, one layer thick. Larvae emerge in about one week and disperse after hatching. Their pale green color makes them very difficult to see on the undersides of leaves. When disturbed, they drop from the leaves suspended by fine silk threads. The larvae are not highly mobile and migrate from an infested field only when the food supply is short, or the crop over-ripe. After feeding, mature larvae burrow 5-15 cm into the soil where they pupate and overwinter. There is one generation per year.

Food Hosts and Damage

Bertha armyworms are general feeders on broadleafed plants, and have been especially harmful to canola and flax on the prairies. They do not feed on cereal crops. Young larvae chew irregular holes in the lower leaves but normally cause little damage. Feeding damage to leaves and pods is first noticeable from early to mid-August. The last two larval stages, the fifth and sixth instars, are the most damaging, with 80 percent of all eaten plant tissue consumed by these stages.

Heavy infestations reduce yields because of defoliation and seed pod consumption. Crop losses due to pod feeding will be most severe if there are few leaves. From a distance, infested canola fields look pale white because larvae eat the outer green layer of the stems and pods exposing white tissue. Larvae can continue to feed on swathed canola for a few days until the crop dries. However, by the time of swathing, most larvae have dropped to the ground to pupate. In some years, early-seeded canola can be swathed prior to damage.

Bertha armyworms have reached outbreak proportions only six times since it was first noted in 1922. Periods of pest abundance have usually lasted 1-3 years and there appears to be no regular sequence to the outbreaks. The large increase in rapeseed/canola acreage contributed to the dramatic population increase in 1971 across the prairies.

Control

Cultivation in the fall and spring will reduce populations of pupae by exposing them to freezing, diseases and predators. Pheromone monitoring of bertha armyworm moths indicates when field sampling is necessary. Bertha armyworms can be controlled with a foliar spray applied when larvae are actively feeding and before the destructive fifth- and sixth-instar larvae appear. The detection of larvae before these stages requires field inspections in late July or early August, depending on when the peak moth flight occurred.

Many diseases and parasites attack the larvae. Viral diseases are common during severe outbreaks, and contribute to sudden declines of armyworm populations. Larvae infected by a virus appear melted onto the plant, while fungus-attacked larvae look as if they have been preserved on the plant. Insect parasites may kill up to 75 percent of the armyworms. Unfortunately, damage to the crop has often been done before the larvae are killed by parasites.

Clover Cutworm
Discestra trifolii (Hufnagel)

Appearance and Life History

The clover cutworm is a periodic crop pest throughout Alberta. Adult moths are uniform or mottled ashy-grey to pale brownish-grey in color, with a wing-span of 25-38 mm. Eggs are white or pale yellow. Young larvae are pale green; older larvae resemble bertha armyworm larvae, but fewer velvety black mature larvae are found and the majority are either green or pale brown. The most distinct difference between the larvae of the two species is that the wide stripe along each side is yellowish-pink on clover cutworms and yellowish-orange on bertha armyworms. Pupae are somewhat smaller than those of the bertha armyworm and have a greenish tinge on the rounded end.

Clover cutworm larvae feeding on canola

H. Philip

There are two generations of clover cutworm per year. They overwinter as pupae in the soil. The first moth generation appears in June, the second in late July or August. Eggs are laid singly on the undersides of leaves in late spring and during the summer. The newly hatched larvae feed on the undersides of lower leaves, gradually moving up the plant as they mature. Mature larvae burrow into the topsoil to pupate. Most of the pupae formed in mid-summer produce the second flight of moths. The pupae formed in August rarely produce a flight of moths; most overwinter to provide moths the following June. The second-generation larvae frequently cause damage at the same time as bertha armyworms (August 10-30), and in years when both are present, clover cutworms may be mistaken for bertha armyworms.

Food Hosts and Damage

Clover cutworms feed on sugarbeets, weeds, forages (especially clover) and various crucifers (mainly canola). They have been an economic problem in localized areas in the Peace River area of Alberta. Clover cutworms feed anywhere on the canola plant and can consume the entire plant. Damage is most noticeable during late June through early July and again from mid-August to September. The economic threshold for this insect is probably similar to that for the bertha armyworm, however, bertha armyworms tend to be dispersed throughout the field, while clover cutworms are found in clumps so that damage is more concentrated. In most years, diseases control the insects resulting in only occasional isolated outbreaks.

Control

Since clover cutworms are climbing cutworms, they can be controlled with foliar insecticidal sprays. Spraying for clover cutworms when canola crops are in flower can put honeybees at risk. Make certain that bee-keepers are informed before spraying.

Diamondback Moth
Plutella xylostella (Linnaeus)

Appearance and Life History

The diamondback moth is a periodic pest of canola across the prairies. The moths are small, 12 mm long, with an 18-20 mm wing-span. They are grey or brownish with white marks on the margin of the forewing. Their name is derived from the series of diamond-shaped figures formed by these white marks when the wings are folded at rest. Minute, disc-shaped, pale green or yellow eggs are laid singly, or in twos or threes, on either side of leaves. A female may lay 30-200 eggs.

Small green larvae hatch within a few days and enter the leaves to feed on, or "mine", the internal leaf tissue. After feeding within the leaf for about a week, the larvae move to the outside of the leaf. The smooth, pale yellowish-green larvae commonly feed on the undersides of leaves. When disturbed, the larvae will move backward rapidly, lashing the body violently, and drop from the plant on a fine silken thread. They will remain hanging several centimetres below the leaf until the danger is past and use the thread to climb back onto the leaf. Larvae feed for 10-30 days, depending upon food supply and temperature, to reach a mature length of about 12 mm.

Diamondback moth larva on canola leaf

Lloyd Harris
Saskatchewan Department of Agriculture

Pupation takes place in delicate, whitish, open lacework cocoons attached on the host or nearby plants. When diamondback moths are extremely numerous, the cocoons may be present in the tens or hundreds on the pods of maturing plants. Adults emerge in 7-14 days; females mate once and egg laying begins almost immediately. Moths do not fly far since they are weak fliers.They feed on nectar from wild flowers at dusk and lay eggs after dark.

There are at least three generations a year, and all stages may be found on the plants at the same time. The first larval generation has only leaves to feed on and is not numerous. Generally, only the second generation larvae cause yield loss when flowering and early podding are at a peak (often about the last week in July). The third generation is likely to affect only unusually late-maturing crops.

These insects do not overwinter on the prairies. Moths are carried into Canada from the southern U.S. on northerly winds in early May or June. The number of spring migrants and their establishment is weather-dependent, and infestations vary greatly from year to year. In the main canola growing areas, most of the crops will not have emerged by the time the moths arrive, so that many eggs are laid on cruciferous weeds and volunteer canola.

Food Hosts and Damage

On the prairies, diamondback moths are most frequently observed in canola fields. Other favored food plants include cabbage, cauliflower, Brussels sprouts, broccoli, turnip and mustard. They do not attack cereals. Damage by young larvae is characterized by small mines and holes or "windows" in the leaves and surface stripping on the undersides of leaves. The amount of leaf area lost probably causes little yield reduction in any but the most extreme cases. Older larvae feed on flowers, young pods and the surface tissue of stems and mature pods, usually from mid-July to early August. Damage seems to be due to feeding on the surface of filling and maturing pods. The seeds under damaged areas do not fill properly and the pods are more susceptible to early shattering. In severe cases, feeding damage shows from a distance as an abnormal whitening. After an infestation is controlled in a podded crop, a new infestation is not likely to develop because of the advancing maturity of the crop.

Control

Factors influencing the potential abundance of this pest are the size of the spring immigration and the availability of suitable food for first-generation larvae. If the moths arrive before preferred host emergence, alternate host plants will include volunteer weeds on summerfallow. Tillage reduces the availability of suitable host plants thus affecting the successful establishment of first-generation larvae. Rainfall is a natural control agent. Young larvae are easily dislodged from plants by rain and can drown on the soil surface or in water trapped on the plants. Other factors influencing diamondback moth abundance are weather conditions during the egg-laying period. Cool, cloudy weather reduces moth flight activity and the longer inclement weather persists, the more females die before egg laying is completed.

Diamondback moths can be controlled by foliar insecticidal sprays. It takes a severe infestation of small larvae to cause appreciable damage. The economic threshold for larvae is 300 per square metre (28 per sq ft).

European Corn Borer
Ostrinia nubilalis (Hübner)

Appearance and Life History

The European corn borer is becoming a serious threat to corn producers across the prairies. Corn borer moths are buff-colored and have a wing-span of about 32 mm. Females are paler, with fewer markings than males. Eggs are flat and, when laid, overlap each other much like fish scales. Just before hatching, the egg mass darkens to the black-head stage because of the darkening head capsules of the larvae. Newly hatched larvae are 3 mm in length, light to tannish-grey in color, and covered with rows of small brown spots. They shed their skins (molt) four times during development, so five larval instars occur. Full-grown larvae are about 2.5 cm in length. Pupae are reddish-brown, about 2.5 cm long, and pointed on one end.

European corn borer larva

Agriculture Canada, Lethbridge

European corn borers overwinter as full-grown larvae in corn stalks, cobs and plant debris on the soil surface. In late spring each larva chews an exit hole and returns to its tunnel to spin a cocoon in which to pupate. Moths begin to emerge in mid-June and fly to hosts, such as tall grass, to find mates and begin egg development. Peak egg laying and flight occur around the middle of July. Females will lay up to 500 white eggs in batches of 15-25 on the undersides of leaves near the midrib. The eggs hatch within a week, and the larvae feed on leaves and work

their way to the whorl of the plant. Pinholes or shotholes are signs of borers already moving into the plant. During later stages of borer development, most larvae enter the tassel, stalk, and ear shank, but they may also feed on the silks, kernels and cobs. Nearly mature larvae bore into the stalk to complete development. Only one generation per year has been observed on the prairies although a second flight has been detected.

Food Hosts and Damage

Sweet corn is the preferred host but grain corn is also damaged. Over 200 other plants including tomatoes, beans, potatoes, oats, sugarbeets and large-stemmed flowers and weeds have been recorded as hosts. All above-ground portions of the corn plant may be attacked. The young larvae bore into growing leaf whorls where they feed on the developing leaves, giving them an etched or shot-hole appearance. Young larvae boring in the leaf midrib will cause leaf breakage. Feeding damage to developing tassel stalks weakens the tassels so they are easily broken off by the wind. Older larvae bore into the stalks and ear shanks, disrupting the normal movement of nutrients and water which then results in reduced yield. Stem breakage and ear drop are common damage symptoms. Boring dust or frass may be observed on leaves, in stems and on kernels within the ear. Feeding damage to sweet corn ears makes them unmarketable. The tunnelling and boring permits secondary infection and damage by stalk- and ear-rotting fungi.

Control

Plowing down corn stubble and residues reduces the number of overwintering larvae by 99 percent and is an effective eradication procedure as long as the strong-flying moths do not migrate into an area to reinfest subsequent crops. Deep plowing in the fall or spring (by May 1) prevents the emerging moths from reaching the soil surface. Corn stalks chopped for silage will remove larvae from the field and destroy them in the ensiling process. Tolerant or resistant corn varieties may be available in the future.

Chemical controls may be effective if applied within a day or so of egg hatching and beginning of larval feeding. Once larvae enter the stalk, no controls are available. Pheromone trapping to determine the time of moth flight and field scouting to determine egg populations and hatching are necessary to ensure timely chemical application. More than one application may be necessary depending on the length of the adult flight and subsequent hatching period.

Pale Western Cutworm

Agrotis orthogonia Morrison

Appearance and Life History

The pale western cutworm has been one of the major economic pests of field crops on the southern prairies for many years. Adults are greyish-white to brownish-white, about 19 mm long with a wing-span of about 38 mm. Several distinct cross lines and two or three spots are found on the forewing. The hind wing is almost entirely white with the outer margins grey or brown. Like most cutworm moths, the body is robust and clothed with long hairy scales. Eggs are spherical in shape and white when first laid but gradually change to dull grey. Newly hatched larvae are about 3 mm long and almost colorless. Fully grown larvae are smooth and lack hair, about 3.8 cm long and greyish-white to green in color with no definite stripes or other markings. The only readily distinguishable characteristic is a yellow-brown head capsule with two distinct vertical black dashes. Pupae are about 2 cm long, brown, shiny and indistinguishable from other cutworm pupae.

Pale western cutworm

Lloyd Harris
Saskatchewan Department of Agriculture

Pale western cutworms overwinter as eggs in the top centimetre of the soil. Hatching takes place in the spring between March and May when the frost has left the ground and the soil is moist. Small larvae can live in the soil for several weeks before starting to feed. They begin feeding on volunteer seedlings and later on seeded crops. Most feeding is done underground with larvae only coming to the surface when the soil is hard or very wet. Larvae pass through 6-8 instars before they become full grown, usually by late June, then they burrow 5-15 cm into the soil to form earthen cells in which to pupate. Adult moths emerge in early August. Moths are very active, feeding on nectar from flowers such as sunflowers and goldenrod. Mating occurs in 2-3 days and egg laying begins almost immediately. Egg laying occurs in late afternoons and early evenings. Eggs are laid just below the surface in loose, friable soil, in cracks in hard packed soil and in stubble fields. Females lay 150-400 eggs each during August and September before frost kills the adults. There is only one generation per year for this native insect.

Food Hosts and Damage

Pale western cutworms are very serious pests of wheat on the prairies. They also attack other cereals, mustard, flax, sugarbeets, legumes and certain weeds such as thistle. The first sign of injury is the appearance of small holes in the leaves. These holes are cut while a portion of the leaf is still underground. As growth continues, the leaf emerges and the holes can be noticed. Cutworms at this time are very small and difficult to find. As the larvae increase in size, they move along the row, cutting off the leaves and often entire plants. Their feeding and resting habits are similar to those of redbacked cutworms. Larger cutworm injury may be recognized by pulling up the plant. If only the plant stems without the roots are removed, cutworm injury is indicated. If the entire plant can be easily pulled, drought may be the cause. Major damage is stem weakening caused by older larvae chewing the stems just below soil level, making the crop susceptible to wind damage. Damage ranges from complete

destruction of individual fields to partial destruction over thousands of hectares. Outbreaks have occurred on the prairies with no apparent pattern since 1911. Precipitation is the critical factor in the increases or decreases of cutworm populations. Too much rainfall in the spring forces the cutworms to the surface where they fall prey to parasites and predators. Very dry conditons result in a decreased food supply, and cutworm death due to desiccation.

Larvae feed under the soil surface, moving up and down with the moisture. Observations and research have determined that 12 or more days of at least 6 mm of rainfall per day is needed during the period from March to June to cause a population decline. If these "wet days" number 10 or less during the March-June period, a population increase is likely to occur. Usually one spring with 12 or more "wet days" will reduce the cutworm population to a point where two or three dry springs are needed to develop a damaging population.

Control

Since this cutworm prefers to lay its eggs in loose soil, fields that are left undisturbed during August and September usually have a protective crust and are much less attractive for egg laying and thus will suffer less damage the following year. Weed-free fields are also less attractive than weedy ones.

In the spring (May) a delay of 5 or more days between cultivation and seeding can prevent infestations because the larvae die if they feed after they hatch and then are deprived of food for several days or if they cannot feed at all for 10-14 days. Although larvae remain below the soil surface, foliar applied insecticides can be effective, because the treated foliage is cut, pulled underground, and consumed by the insect. Contact or vaporization of the insecticide may also contribute to cutworm death. Since these cutworms feed at night, spraying should be done in the evening. Economic thresholds are similar to those of redbacked cutworms.

Redbacked Cutworm
Euxoa ochrogaster (Guenée)

Appearance and Life History

The redbacked cutworm occurs throughout Canada and the northern United States and is capable of causing extensive damage even at low population densities. Adult moths are light fawn to brick-red in color with a wing-span of about 40 mm. There are four color variations of redbacked cutworm adults. The wing pattern is similar but the distinguishing features may be unclear. Males are brighter than the females. Eggs are tiny, globular and white and very difficult to find. Newly hatched larvae are about 3 mm long but grow quickly. The larvae molt five or six times during their life cycle and mature as dull grey, hairless caterpillars with a red or reddish-brown top stripe, usually extending the entire length of the body. The top stripe is divided by a dark line and bordered with darker bands. The head is a yellowish-brown. Mature larvae may measure up to 38 mm long. Pupae are reddish-brown in color and similar in size and appearance to other cutworm species.

Redbacked cutworm larvae

M. Herbut

Redbacked cutworms overwinter in the egg stage. Eggs usually hatch in April as soil temperatures increase. Larvae begin feeding immediately on any nearby plants and feed for 6-8 weeks with most of the damage apparent in June. Larvae generally remain inactive during the day, but during the night, either come to the surface or move underground in search of food plants. When mature, the larvae pupate in an earthen cell 2-5 cm in the soil. Adults begin emerging in late July and continue emerging until late August. Adults are nocturnal and most active on warm nights. Mating takes place within 1 or 2 days after emergence and egg laying begins a few days later. Females are capable of laying about 400 eggs over a 20-day period. Eggs are laid just below the soil surface in cultivated fields, especially in loose, dry soil in weedy stubble or fallow fields. They prefer the light to medium type soils found in the more humid, northern regions of the prairies. There is one generation each year.

Food Host and Damage

The primary food hosts are cereals, sugarbeet, flax, canola and mustard. Redbacked cutworms have been recorded feeding on most vegetables, sunflower, sweetclover, alsike, alfalfa, various tree seedlings and garden flowers. Damage by young larvae is characterized by small holes and notches in the foliage. Older larvae eat into the stems and usually sever them at or just below the soil surface. Infestations in cereal crops are characterized by areas of bare soil that gradually enlarge until anywhere from 1 to 2 acres to complete fields of grain are destroyed. These bare areas of exposed soil are often confused with areas of poor germination or moisture stress. The presence of cutworms is indicated by dead, dried up plants that have been severed.

Infestations last 2-4 years followed by a minimum of 2 years of relative scarcity. Factors contributing to population increases are hot, dry conditions in August allowing greater moth feeding activity which is critical for egg production and optimum egg laying, readily available egg-laying sites, availability of suitable flowers to provide nectar for the moths, and low parasite populations. During the first year of an outbreak very few parasites are found. However, after 2 years, the parasites are numerous enough to reduce the outbreak and keep cutworm numbers low for at least 2 years, depending on the factors already mentioned.

Control

Crusted soils from late July until late September on summerfallow help prevent egg laying. This is earlier than pale western cutworm (August-September). If weed growth develops in August, it should be destroyed as redbacked cutworm moths usually lay their eggs in weedy summerfallow. They also lay in weedy patches in cereal crops and in fields of canola, peas, alfalfa and sweetclover. Young cutworm larvae may be starved before spring seeding by allowing volunteer growth to reach 3-5 cm, cultivating, and then seeding 10-14 days later. Cutworm infestations can be confirmed by digging about 3 cm below the soil surface in the damaged area and finding cutworms. More than 1 cutworm per 30 cm of crop row may require insecticide application. Only infested areas of the field need to be sprayed. Since cutworms feed only at night, spraying should be done in the evening.

Sunflower Moth
Homoeosoma electellum (Hulst)

Appearance and Life History

The adult sunflower moth is shiny grey to greyish-tan, about 9 mm long with a wing-span of about 20 mm. The forewings have a small dark dot near the centre of each wing and two or three small dark dots near the leading edge of each wing. The hind wings are plain. At rest, the wings are held tightly along the body giving the moth a somewhat cigar-shaped appearance. Eggs are oval, pearly white, and just visible to the naked eye. Larvae are purplish- or reddish-brown with four longitudinal, light grey stripes on their backs and measure about 20 mm long when full grown. Pupae are reddish-yellow and about 20 mm long.

Sunflower moths do not overwinter on the prairies but migrate north from the south central United States. Moths may appear on the prairies sometime in late June or early July and are attracted to sunflowers that are beginning to bloom. Eggs are laid on and in the florets of the newly opened sunflower head where pollen is abundant. A female may lay up

Sunflower moth larva feeding in sunflower head

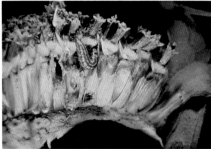

Agriculture Canada, Winnipeg

to 30 eggs per day. Eggs hatch in 2-4 days and the newly emerged larvae feed on pollen and florets. Upon reaching the third larval growth stage, the larvae begin tunnelling into seeds. This tunnelling and feeding continues throughout the remainder of the larval development (about 1 month). Mature larvae lower themselves to the ground on strands of silk and enter the soil where they spin cocoons in which to pupate. A second flight of moths occurs in early September but it is difficult for a second generation of larvae to mature before winter.

Food Hosts and Damage

Sunflower moths are occasionally severe pests of sunflowers on the prairies. They feed mainly on wild and cultivated sunflowers but have also damaged zinnias, daisies and chrysanthemums. Young larvae feed chiefly on the pollen and florets, but as the larvae develop they tunnel into and through the seeds and head tissue. While feeding, the larvae spin silk threads, which become matted with florets and frass and give the head a trashy appearance. Each larva is estimated to destroy an average of nine seeds but some may destroy up to 25. The percentage of seeds destroyed per head is equal to the percentage loss in yield of the harvested crop. Infestations may become sufficiently heavy to destroy all the seeds in a head. Early-seeded sunflower crops seem to be more susceptible to damage than late-seeded crops.

Control

Since sunflower moths move northward on air currents, they are not a problem every year and controls are not always necessary. Scouting fields for adult moths when sunflowers first come into bloom is necessary to determine the presence and density of moths. Observations should be made in the early morning or early evening when the moths are most active. Treatment is recommended if one or more moths are found per five plants at the onset of bloom or within 7 days of the first appearance of adults. Fields that are in bloom or that bloom 2 weeks or more after the first adult moth appearance have very low potential for damage despite the presence of moths. Insecticides must be applied while the flowers are in bloom, killing the moths before the eggs can be laid to prevent the larvae from hatching and tunnelling into the head where the insecticide does not penetrate. Insecticidal applications should be done in the evening when the moths are most active and pollinators are least active.

Orthoptera
Clearwinged Grasshopper

Camnula pellucida (Scudder)

Appearance and Life History

The clearwinged or "roadside" grasshopper is the second most destructive species of grasshopper in western Canada. Adults are small- to medium-sized grasshoppers (males 23 mm, females 30 mm) with color varying from pale yellowish-brown to black with a markedly lighter undersurface. When folded, the forewings have two pale stripes beginning at the thorax and converging at the tip of the forewings. Round or oval dark blotches are also seen on the forewings. The hind wings, when expanded, are clear with no markings. The hind legs are pale with three narrow dark bars. Young nymphs are dark to almost black with a distinctive white band encircling the thorax. Older nymphs are mottled grey or brown. Adults lay 20-30 light brown eggs, 4.5 mm in length and 1.5 mm in diameter,

arranged in two columns in each egg pod.

Adult clearwinged grasshopper

Agriculture Canada, Lethbridge

Clearwinged grasshoppers overwinter as eggs. Hatching occurs in late May or early June with nymphs feeding for 4-8 weeks and molting five times. Adults appear in late July or early August. They gather on the margins of unplowed land, pastures and ditch banks to feed and mate. Egg pods are laid in sod in permanent egg beds usually on or near hilltops or knolls and roadsides. An egg pod is laid about every 10 days from early August until a killing frost. Upwards of 10 000 hoppers per square metre may be found hatching during outbreak years. Flight is local since females return to the egg beds at night and at egg-laying time.

Food Hosts and Damage

Clearwinged grasshoppers prefer sedges and grasses, both native and introduced, as food and rarely eat broadleafed plants. Lush cereal crops are especially attractive. Damage is primarily along margins of crops. Clearwinged grasshoppers will often chew crops and grasses right to the ground. They are especially damaging in the nymphal stage as many eggs hatch at the same time and large numbers consume all the green vegetation available.

Control

Controls are similar to those for migratory grasshoppers. Indentification of egg beds and spraying when hoppers are still small will help reduce damage.

Migratory Grasshopper
Melanoplus sanquinipes (Fabricius)

Appearance and Life History

The migratory grasshopper is the most widespread and most destructive grasshopper in western Canada. Adults are greyish-yellow to brown in color. The hind legs are marked with a series of black bands and a pink stripe along the lower edge. A black band extends along the side of the thorax just behind the eyes. The males are about 20 mm and the females about 28 mm long. Yellowish eggs, 4.8 mm long and 1 mm in diameter, are laid in cylindrical pods containing 12-20 eggs arranged in two columns. Newly hatched hoppers closely resemble adults but are wingless and less than 6 mm long. There are usually five nymphal instars which can be recognized by the mottled black appearance and the black bands along the top of the thorax.

Adult migratory grasshopper

Agriculture Canada, Lethbridge

Migratory grasshoppers overwinter as eggs in the soil. Hatching occurs from early May to mid-July, making it one of the first species to hatch in the spring and, depending on temperature and moisture conditions, one with the longest hatching period. Nymphs feed for about a month before reaching the winged adult stage. Egg laying begins about a week after the female reaches the adult stage. Females lay 200-300 eggs from late July into the fall. Eggs are laid at 2- to 3-day intervals in pods 5 cm deep in open soil such as soil drift or among plant crowns in weedy pastures, roadside ditches and stubble fields. Moist sandy soil is preferred over loam and clay soils or dry sandy soils. The total number of eggs laid depends on temperature, food quantity, and vitality of the female. There is one generation per year.

Migratory grasshoppers get their name from the mass flights that have occurred during outbreak years. These flights are generally downwind on warm southerly winds. Feeding and activity occur mainly between 20 and 32°C. Grasshoppers rest and tend to roost on plants when the temperature is below or above that range.

Food Hosts and Damage

Migratory grasshoppers are mixed feeders on both grass and broadleafed plant material. They thrive in weedy grain fields, cultivated pastures and hay fields, roadside ditches and garden crops. They feed on the leaves, stems, flowers, fruits and seeds. Damage to cereal crops is generally concentrated near field margins as they move from adjacent fields and waste areas. Grasshoppers that hatch in crops seeded on stubble fields feed on growing seedlings and damage may go unnoticed until extensive leaf chewing has taken place. Besides feeding on leaves of young plants, grasshoppers may clip the stems just below the seed heads of the crop, causing very heavy losses. Crop damage is directly related to population densities.

Population levels are primarily determined by weather conditions. Outbreaks are usually preceded by 2-3 years of hot, dry summers and open falls. Dry weather not only increases egg and nymphal development but also slows down plant growth. Open falls allow grasshoppers more time to feed and lay eggs and allow more complete egg development for faster and more even hatching the next spring. Cool, wet weather increases the probability of egg and nymphal mortality due to diseases, slows grasshopper growth and activity, and helps the crop overcome stress.

Control

Grasshoppers are attacked by several species of parasites and pathogens which along with unfavorable weather assist in reducing outbreaks, especially in localized areas. Cultural and preventive measures for grasshopper control include early seeding of crops to give the crops a head start, crop

rotations to avoid planting cereals on stubble fields heavily infested with grasshoppers, tillage to remove food hosts since the nymphs are wingless and do not move very far, and trap strips to collect grasshoppers into a small area for chemical treatment.

When cultural methods are not practical or the populations are too high, a number of registered insecticides can be used to control grasshoppers. Applications should be directed against the small nymphs which require lower treatment rates than the larger adults.

Twostriped Grasshopper
Melanoplus bivittatus (Say)

Appearance and Life History

The twostriped grasshopper is one of the largest and most conspicuous of the cropland grasshoppers. Males are about 24-28 mm long whereas females may be up to 40 mm in length. Adults are dark yellowish-green in color with a pair of distinct yellow stripes extending from the side of the head behind the eyes to the tip of the forewings. A solid black stripe is evident along the outer side of the yellowish hind legs. Nymphs are yellowish but have the black stripe on at least a third of the hind leg. Eggs are green to yellow-brown in color, 4.4 mm long and 1 mm in diameter; 40-100 eggs are arranged in four columns within each egg pod.

Adult twostriped grasshopper

Agriculture Canada, Lethbridge

Twostriped grasshoppers overwinter as eggs that hatch in late May to early June. Nymphs feed for 5-6 weeks and pass through five nymphal stages. Adults appear in the last 2 weeks of July and lay eggs in heavier textured soils along roadsides, closely cropped pastures, fence rows, ditch banks, prairie sod and field margins but not in cultivated fields. Temperatures must be above 20°C and soil moisture 10-20 percent for egg laying to occur. Only two or three pods are laid by each female during August and September. There is only one generation per year.

Food Hosts and Damage

Twostriped grasshoppers feed on both grasses and broadleafed plants, with the latter being necessary for maximum growth. They prefer the lush growth found around edges of streams, marshes and cultivated fields. Hosts include weeds, most crops, especially alfalfa and vegetables, and occasionally trees and shrubs. Individually, they are the largest feeders and most destructive of the prairie grasshoppers. Their numbers, however, do not usually become as great as the migratory grasshopper, nor do they swarm and move great distances. Environmental conditions favoring population increases are similar to the migratory grasshopper.

Control

Controls for the twostriped grasshopper are similar to those for the migratory grasshopper. Spraying roadside ditches and field margins are effective controls for this species.

Thysanoptera
Barley Thrips
Limothrips denticornis (Haliday)

Appearance and Life History
The barley thrips, an introduced European species, is a minor but common pest of barley in most areas of the prairies. Adult thrips are dark brown to blackish and about 2 mm long. The female has wings fringed with long hairs and has prominent bristles at the end of her pointed abdomen. The wingless male has a rounded end to the abdomen. The wingless, immature forms are pale greenish-yellow and blend in with the host plant making them difficult to see. Barley thrips may be separated from other thrips by an angular projection on the third antennal segment.

Female barley thrips emerge from their overwintering sites in late May and early June and may be found on grasses in and around shelterbelts. They feed within the leaf sheaths of grasses and early seeded barley. Females crawl down inside the terminal leaf sheaths where they lay about 15 eggs in the soft leaf tissue. Eggs hatch in 4 or 5 days in late June. Larvae are usually confined to the inner surfaces of the terminal leaf sheaths. In this species it is the female prepupa (immature stage) and not the adult which is fertilized by the male adult, a most unusual practice. Development from egg to adult takes about 3 weeks when the new generation appears. The males die off and only the females overwinter. They prefer to hibernate in bromegrass or bluegrass sod along shelterbelts. Overwintering mortality is about 60-70 percent. Mass migrations occur in late July as they move to new hosts. Barley thrips may also be blown in from the U.S. each spring.

Food Hosts and Damage

The summer generations prefer barley to other grasses or cereals. They may be found feeding both within the terminal sheath or within the barley head, just as the head is emerging from the boot. The thrips work themselves under the glumes

Barley head damaged by barley thrips

H. Philip

and pierce and suck the plant juices of both the glume and the developing grain kernel. Losses are generally confined to a reduction in test weight and an increase in thin or smaller kernels. Blind seeds may be compensated for by the remaining larger and plumper kernels.

The presence of "white heads" in the field also indicates thrips damage, however, other agents have been implicated with this condition. Fields should be checked when the heads are just beginning to emerge from the sheaths and before the crop is fully headed. Peel the upper sheath and look for the dark adults on the leaves and developing head.

Control

Spraying for barley thrips is rarely necessary. No definite economic thresholds have been determined but reports from North Dakota state that 30-60 thrips per head caused no economic damage. This is the offspring of 2-4 females per stem. The only time an insecticide treatment would be of benefit is when applied as the heads emerge from the boot.

Insect Pests of Small Fruits and Vegetables

Coleoptera
Colorado Potato Beetle
Leptinotarsa decemlineata (Say)

Appearance and Life History

The Colorado potato beetle, a native of the high plains of the eastern Rocky Mountain slopes of the United States, is now found in every province of Canada. Adults are 6-12 mm long, oval, convex, and yellowish with black markings on the head and shoulders, and ten black longitudinal stripes along the back. Orange eggs are elongate and arranged in clusters of ten or more on the undersides of leaves of host plants. Young dark larvae develop into orange or red, soft, hump-backed larvae which have two rows of dark spots along each side. The head and feet are both red. They are about the same size as the adults. Pupae are orange, soft-bodied and resemble adults in form.

Colorado potato beetle larva feeding on potato

M. Herbut

Colorado potato beetles hibernate as adults in the soil at depths to about 40 cm, depending on the climate and soil type. Adults emerge in May or June and immediately fly in search of food hosts. Females mate, lay up to 500 eggs and then die. Eggs are laid on the undersides of leaves and hatch in 5-10 days. Larvae pass through four growth stages in 2-3 weeks, after which they drop to the ground and pupate 5-10 cm beneath the soil surface. Adult beetles emerge 5-10 days later. Depending on weather conditions and food availability, the number of generations produced per year varies from one to two and a partial third.

The number of adults which survive the winter determines the size of an infestation. Abundant snow fall increases the survival rate by decreasing mortality due to freezing. Cool wet weather during the egg-laying period can reduce population levels by reducing adult activity and increasing egg mortality.

Food Hosts and Damage

Colorado potato beetles can be serious pests of potatoes wherever they are grown. They will also feed on tomato, eggplant and wild tomato. Plants can be completely defoliated in a short period of time since both larvae and adults feed on the same plant. Larvae, however, do the greatest damage. Extensive feeding on the leaves and stem tips prevents tuber development in potatoes or fruit development in other host plants. These beetles also transmit plant pathogens that cause such potato diseases as spindle tuber, bacterial wilt and bacterial ring-rot.

Control

There are no practical controls to prevent Colorado potato beetles from flying into a crop to lay eggs. Chemical controls may be required if more than two larvae are found per young plant or five per larger plant. Controls after flowering may not be economical if tuber development is progressing well. Hand picking adults or larvae is usually all the control required in home gardens. Predation and parasitism have little effect on the development of infestations.

Spinach Carrion Beetle
Aclypea bituberosa (LeConte)

Appearance and Life History

The spinach carrion beetle is believed to be an introduced insect which feeds on plants in the goosefoot family and on carrion. Adult beetles are 10-12 mm in length, dull black, oval, somewhat flattened, and have elevated ridges running lengthwise on the wing covers. Eggs are oval and creamy white with a polished and glistening surface, and measure about 1.5 mm in length. Larvae are shiny black, 6-12 mm long, wedge shaped, and flattened with distinct body segments. Pupae look very similar to adults but are pure white, soft, and about 12 mm long.

Spinach carrion beetle larva

M. Herbut

Adult spinach carrion beetles overwinter in the soil along field margins, ditch banks and roadsides. In May and early June mated females lay their eggs in the soil to a depth of 5 cm. Eggs hatch in about a week and the young larvae start feeding immediately. The larvae are easily disturbed and, when frightened, tumble to the ground and crawl rapidly under clods of soil. Larvae prefer to feed at night, but when they are abundant in a field or garden, some may feed during the day. When full grown, they burrow into the soil to a depth of 2-5 cm and construct oval cells in which to pupate. Adults emerge from the soil in 2-4 weeks and commence feeding on host plants. They will seek shelter under clods of soil when disturbed.

Food Hosts and Damage

Larvae feed on the leaves of various plants, notably sugarbeet, spinach, pumpkin and squash. Other crops attacked include cabbage, radish, rhubarb, potato, lettuce, turnip, pea, Swiss chard and strawberry. Weeds attacked include spear-leaved goosefoot, lamb's quarter and redroot pigweed. Adults and larvae eat leaves; the greatest damage occurs in May while the plants are small. Ragged edges and large holes in the leaves indicate their presence. They are often more abundant along field borders. Entire rows of spinach or cabbage may be destroyed in the home garden.

Control

As these insects commonly migrate from weeds onto the crop, removal of weeds such as lamb's quarters and redroot pigweed from field borders and gardens will usually prevent populations from building up. Chemical controls are available if damage to crops is expected because of large numbers of larvae. Chemicals should be applied in the spring when the crops are still small and the larvae are feeding. Control of the adults is rarely necessary.

Tuber Flea Beetle

Epitrix tuberis Gentner

Appearance and Life History

The tuber flea beetle is a recent introduction to the prairies and poses a threat to potato production. Adults are oval, dull black beetles with reddish antennae and legs. They measure 1.7-2 mm in length and are covered with fine hairs visible with a magnifying glass. The femur of the hind leg is greatly enlarged, allowing the beetles to jump considerable distances relative to their size. Eggs are very small. Larvae are small, less than 5 mm in length, whitish, cylindrical grubs. Pupae resemble the adult but are white to brownish in color and soft-bodied.

Adults overwinter in the soil in protected places and emerge in the spring, usually about mid-June on the northern prairies. They feed on weeds and other suitable vegetation until the preferred garden

plants are available. Females lay about 90 eggs in the soil around the base of the plants. Eggs hatch in about a week. Larval development takes place in the tuber and may take 3-4 weeks to complete. The pupal stage is completed in 7-10 days. Adults emerge in mid-summer and feed on the leaves and then migrate to overwintering sites in the fall. Hot dry summers are favorable and promote larval survival and population build up. Severe winter cold contributes to heavy mortality. It is believed there is only one generation a year, however, there may be two in favored locations on the southern prairies.

Food Hosts and Damage

The main host for the tuber flea beetle is potato but they also feed on tomato, pepper, eggplant, bean, cabbage, corn, cucumber, lettuce, radish, spinach and various weeds when potatoes are not available. Adult beetles chew small round holes in the leaves, producing the common "shot-hole" damage typical of many flea beetles. Larvae feed below ground on roots and tubers. Feeding damage to the tubers is characterized by winding grooves on the surface, or by narrow tunnels extending up to 6 mm into the tuber which become filled with brown corky material. Before peeling, tubers have a "pimply" appearance which may be mistaken for common scab or wart disease. With extra peeling, the potatoes are perfectly safe to eat. Commercial marketability of damaged tubers is greatly reduced. Larval feeding also allows entry of many microorganisms which may cause rotting and/or reduce the storage potential of the potato. Russet Burbank (Netted Gem) potatoes are more severely affected than Warba, Kennebec and Pontiac.

Control

Control measures are aimed at killing adult beetles before they lay eggs. Insecticides must be applied when beetles or leaf shot-holes are first observed and then repeated at 7- to 10-day intervals as long as beetles are present. Growers may consider early planting of a few potatoes as a trap crop in the same area as the previous year's

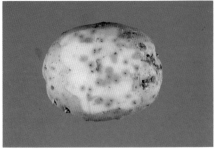

Potato tuber damaged by tuber flea beetle larvae

M. Herbut

crop. Locate current crop as far as possible from previous year's crop and away from urban areas. Destroy volunteer plants and cull potatoes to reduce population development.

Diptera
Beet Leafminer

Pegomya hyoscyami (Panzer)

Appearance and Life History

The beet leafminer is present wherever sugarbeets are grown in North America. Adults are grey flies about 7 mm long and resemble the common housefly. Their bodies are covered with long, stiff bristle-like hairs, and their abdomens are bent downward and curved under. Eggs are white, elongate and about 1 mm long. They should not be confused with the flat scale-like eggs of the beet webworm which may be present at the same time on the undersides of the beet leaves. Larvae, which are found inside the leaf, are white or yellowish, without head or legs, and tapered from front to back. As it is almost transparent, the green leaf tissue it consumes may be seen within the body. Puparia are oval, brown, and about 5 mm long.

Beet leafminers overwinter as pupae in the soil beneath host plants. In late May the adults emerge and crawl to the surface of the soil. After mating, the females place their eggs singly or in rows of 2-10 on the undersides of the leaves of host plants. After hatching from the eggs, the tiny larvae eat holes through the surface of the leaf and feed between the upper and lower surfaces of the leaf. When the maggots are full grown, they drop to the soil and crawl beneath the soil surface to a depth of 5-8 cm, where they change into pupae. There is usually more than one generation a year. This insect is also known as the spinach leafminer.

Food Hosts and Damage

Sugarbeets, spinach, Swiss chard and beets are damaged by the larvae mining inside the leaves. Several weeds such as lamb's quarters also serve as hosts. The newly hatched larvae make holes in the outer covering of a leaf and feed on the inner tissue making narrow thread-like mines. Mature larvae produce large clear irregular blotches or blisters 12-15 mm in diameter. The destruction of the inner part of the leaf has the same effect as destruction of the entire leaf. Larval mines make the leaves unfit for eating, canning, or sale. Damage to young sugarbeets has always been light, but older plants are quite often damaged. As the older leaves are the ones most frequently infested, and generally shaded by the newer leaves late in the season, they do not make a very significant photosynthetic contribution to the plant.

Beet leaf damaged by the beet leafminer

M. Herbut

Control

Leafminers are attacked throughout the season by parasites, predators, and diseases which tend to keep them under control. Destroying lamb's quarters and other weeds in fields and ditches helps prevent the pest from becoming established on its weed hosts. Planting early, maintaining high soil fertility, and providing adequate soil moisture enables plants to resist the injurious effects of an infestation.

Insecticides may be used as soon as the mines become apparent. However, control of the larvae mining inside the leaf is very difficult.

Cabbage Maggot
Delia radicum (Linnaeus)

Appearance and Life History

The cabbage maggot is a pest of cruciferous crops across the prairies. Adults look similar to house flies but are smaller (5 mm long), dark ash grey in color with a dark stripe along the top of the abdomen, and covered with black bristles. The reddish purple eyes on males nearly touch in the center of the head while female eyes are separated. Eggs are white, elongated and about 1 mm long. Larvae are small, white, legless maggots that reach a length of 6-10 mm when full grown. Puparia are elongate, brown, have rounded ends, and resemble small wheat kernels.

Cabbage root maggots feeding on cabbage root

M. Herbut

Adults are found from late spring to late October flying close to the ground in search of suitable host plants. Flies emerge in the spring from overwintering puparia and feed on the nectar of wild flowers. They mate and females begin egg laying about a week after emerging. During their 5- to 6-week life span, females lay 50-200 eggs singly or in masses at or near the stems of host plants on cool, moist soil. Depending on temperature, maggots hatch in 3-10 days and commence feeding on small roots and root hairs and then tunnel into the main roots. They mature in about 3-4 weeks, then leave the roots and pupate in puparia about 5-20 cm deep in the soil. Adult flies emerge in 2-3 weeks, mate, lay eggs, and repeat the cycle. The

number of generations produced annually varies from one to two and sometimes a partial third. Generations tend to overlap; thus all stages of the life cycle can be found throughout the summer months.

Food Hosts and Damage

As the name cabbage maggot implies, this insect can be a very serious pest of cabbage on the prairies. Other plants attacked include rutabaga, cauliflower, broccoli, Brussels sprouts, radish, turnip, canola and wild mustard. The first generation of maggots often destroys seedlings during the first part of the growing season by girdling the roots. The outer leaves turn yellow and wilt indicating that maggots are present. The second generation does not usually kill the host plants but will reduce marketable yields, especially of cole root crops. Feeding injury allows entry of root rot organisms. Vegetables infested with maggots appear pale green, stunted, and will wilt on hot days. Maggot damage may cause canola crops to lodge. Moisture conditions and soil type influence the amount of loss due to maggot infestation. Cool, wet conditions promote infestations; hot, dry conditions reduce or delay infestations. However, wet weather encourages rapid plant growth which in turn reduces damage to the plants. In general, maggots do more damage to plants grown in light soils than in heavy soils. Another species that attacks the same crops and has a similar life cycle is the turnip maggot, Delia floralis (Fallen).

Control

Several species of parasites and predators attack both of these species of root maggots. Removing all cruciferous weeds from the garden and placing aluminum or tar paper disks with a 15 cm diameter around the plant at ground level gives some protection to nonroot crops. A granular application of an insecticide in the furrow to direct-seeded crucifers will aid in controlling first-generation maggots. A soil drench of an insecticide at the time of transplanting and again at 2-week intervals until the end of June and again from late July to mid-August will aid in reducing root damage. Since eggs hatch within 2 or 3 days after being laid, it is important that the insecticide be applied promptly at the recommended time.

Currant Fruit Fly

Epochra canadensis (Loew)

Appearance and Life History

The adult currant fruit fly is a small yellowish insect about half the size of a house fly with smoky or dark bands across its wings and bright green eyes. Eggs are shiny white, about 1 mm long, and have a small stalk at one end. Larvae, or maggots, are whitish, legless, and may reach a length of about 7 mm. Puparia are about the size of a wheat kernel, smooth, and light to dark brown in color.

Currant fruit flies overwinter as pupae in the soil beneath host plants. Flies emerge about the time their food hosts are in full flower and may be found resting on the underside of the leaves. After mating, the females will lay 100-200 eggs, one at a time, in developing berries. Eggs hatch in about a week and the maggots feed on the developing fruit tissue. The maggots mature about the time the fruit is ripe, causing the fruit to drop to the ground. Maggots leave the fruit several days later and enter the soil beneath the leaf litter. They pupate within the top 5 cm of the soil and remain there until next spring. There is only one generation per year.

Food Hosts and Damage

Red and white currants and gooseberries are the main hosts of these native insects in western Canada. They cause damage wherever currants or gooseberries are grown. Infested fruits are discolored with a red spot where the eggs are inserted. The feeding action of the maggots may cause the berries to drop prematurely, thus reducing berry yields. Infested berries that are harvested are "wormy" and must be culled.

Currant fruit fly puparia, larvae and feeding damage to gooseberry

H. Philip

Control

Raking the soil under bushes in the summer to remove infested berries and tilling the soil 5-7 cm may help in reducing the overwintering population. Properly timed insecticidal sprays on the undersides of leaves when the adults first emerge and again about a week later will give good control. Sprays can also be applied when the flowers are withering or during petal drop.

Onion Maggot

Delia antiqua (Meigen)

Appearance and Life History

The onion maggot is the most important insect pest of onions on the prairies. Adult flies resemble houseflies but are smaller (6 mm long). They are greyish-black in color, have brown stripes on the thorax, and are covered with short, bristly hairs. Eggs are white and elongated, about 1.25 mm long. The legless larvae or maggots, which grow to a length of 8 mm, are creamy white in color with tiny black mouth hooks. Chestnut-brown puparia are 4-7 mm long.

Onion infested with onion root maggots

H. Philip

Onion maggots overwinter as pupae in the soil. Adult flies emerge from mid-May to late June, mate and begin egg laying about 2 weeks later. Eggs are laid either in the soil near host plants or on the stems and leaves of host plants. Each female is capable of laying several hundred eggs. Eggs hatch in 2-6 days and the small maggots begin feeding on host plants below the soil surface. After 2-5 weeks of feeding, the maggots leave the plants and pupate in the soil 5-12 mm below ground level. Adults emerge in 2-3 weeks and the cycle is repeated. On the prairies there are one to three full generations per year depending on soil and weather conditions.

Food Hosts and Damage

Onion maggots attack onion, bunching onion, garlic, leek and chives, with onions being the preferred host and most seriously damaged. The first sign of injury is the wilting of foliage. The whole plant becomes flaccid and then collapses. This damage is particularly noticeable in the seedling stage. Maggots frequently destroy groups of plants which leads to patchy crops. Larger onions survive the attack but distorted growth accompanied by rotting of the tissues makes them unmarketable. Onion bulbs infested just prior to harvest will continue to rot while in storage.

Onion maggot infestations are greatly influenced by soil, temperature and weather conditions. Infestations occur more often and cause greater economic damage in irrigated low-lying areas. Onions sown in newly ploughed soils are very susceptible to severe infestations. Onion maggots are more abundant during wet years and a succession of wet years will result in population increase with a subsequent increase in potential losses due to onion maggot damage. A long summer season with an open fall may allow development of a third generation of larvae that will attack mature and harvestable onions.

Control

The eggs, larvae and pupae of onion maggots are readily attacked by several predators and parasites. Crop rotations help to reduce onion maggot populations but adult flies will move from yard to yard during second- or third-generation flights to reinfest gardens. Cull onions and onion refuse should be destroyed to remove any overwintering sites. Chemicals may be applied at seeding as furrow applications to control first-generation maggots, or as sprays to control adults in the spring, or during the summer when adults of the second generation are in flight, or as soil drenches in late June to control second-generation maggots.

Homoptera
Aster Leafhopper
Macrosteles fascifrons (Stål)

Appearance and Life History

The aster leafhopper is an important vector of the viruses that cause aster yellows and other viral diseases of many crops. Adults are 3.5-4 mm long, light green to yellowish-green in color, and have six black spots arranged in pairs on the front of the head. This species is often called the six-spotted leafhopper. The forewings are smokey and held roof-like over the back. Adults jump and fly readily. Nymphs range from 0.6 to 3 mm in length and have similar markings to the adult but the color varies from yellow or light brown to a pale greenish-grey. Nymphs lack wings but are also active.

Adult aster leafhopper

M. Herbut

Aster leafhoppers overwinter in the southern U.S. and migrate northward each spring, arriving in great numbers from mid-May to mid-June. Eggs are laid in plant tissue and the nymphs hatch within a few days and commence feeding on host plants. As many as four generations may be produced per year on the prairies.

Food Hosts and Damage

Aster leafhoppers feed first on fall rye and wheatgrass, then on spring grasses and flax, and finally on fall rye again. They also feed on lettuce, celery, carrot, potato, sunflower, parsnip and oats and barley. They feed by piercing plant tissue and sucking up the plant fluids. Leafhopper feeding can cause discoloration, wrinkling, and wilting or "burning" of leaves and dwarfing of the plants. A wide range of yield-reducing yellows and other viruses and virus-like pathogens are transmitted from diseased plants to healthy plants by these insects. An incubation period of 10-18 days is required before a leafhopper is capable of transmitting the yellows virus to other plants.

Control

Controls are rarely necessary except in crops being grown for seed. Chemicals applied at the time the adults are preparing for egg laying is the best time to spray.

Hymenoptera
Imported Currantworm
Nematus ribesii (Scopoli)

Appearance and Life History

The adult imported currantworm is a sawfly measuring about 8 mm long, and black with pale yellow lines on the abdomen. Eggs are white, shiny and elongate. Larvae pass through two color phases before pupation. Immature currantworms are yellowish-green with black head and legs and many black spots on the body. Mature larvae are 20 mm long without any black on them. They are uniform green in color with yellowish heads and terminal segments. Larvae have three pairs of legs on the thorax

and eight pairs of false legs (prolegs) on the abdomen. If disturbed, the larvae raise the front and rear ends of their bodies. The brown pupa is inside a silken cocoon.

Immature (left) and mature (right) imported currantworm larvae

M. Herbut

Imported currantworms pass the winter on or near the soil surface in cocoons as either larvae or pupae. Adults emerge in May and June and begin egg laying. Eggs are attached in rows to the veins and midribs on the undersides of host plant leaves shortly before the plant comes into full foliage. Eggs hatch in 7-10 days. After 2 or 3 weeks of feeding, the larvae mature and pupate.

The second brood of adults emerges in late June or July. Larvae again appear in July. These second-generation larvae overwinter in cocoons on or near the soil surface.

Food Hosts and Damage

Imported currantworms feed exclusively on wild and cultivated currant and gooseberry. Black currant is not usually attacked. Larvae devour leaves and can completely defoliate a bush, thus reducing berry yields and weakening the plants. Damage usually starts in the thick foliage near the centre of the plant and moves outward. This pest is found wherever currants or gooseberries are grown.

Control

Larvae can be readily controlled with an insecticide when they are first noticed. Hand-picking is also effective if the larvae are not numerous or few bushes are affected.

Raspberry Sawfly
Monophadnoides geniculatus (Hartig)

Appearance and Life History

The adult raspberry sawfly is about 6 mm long, black with yellow and reddish markings, and has four clear wings. Mature larvae measure 12 mm long, are light green in color, with all segments except the head possessing a number of conspicuous whitish bristles.

Raspberry sawfly larva

M. Herbut

Adults emerge in the spring about the time raspberry is in blossom and insert their eggs into leaf tissue. Larvae hatch and feed on the underside of the leaves during the summer, and then drop to the ground to construct a cocoon in the soil in which to overwinter and pupate the next spring.

Food Hosts and Damage

Raspberry sawfly larvae feed mainly on raspberry leaves but may be found on other cane fruit. Larvae may also feed on flower buds, young fruit and tender bark of growing shoots. Initial damage to the leaves appears as small holes.The larvae will completely devour the leaves, leaving characteristic elongated holes between the larger veins. Defoliation results in weakened plants and reduced berry yields. The raspberry sawfly is usually not a serious pest but sometimes is numerous enough to attract attention because of the skeletonized raspberry foliage.

Control

Damage is not usually severe enough to require control. Vigorous raspberry plants are not seriously damaged by sawfly larvae unless they are present in outbreak numbers. Insecticides may be applied in spring before the first blossoms open to control adults or may be applied later to the undersides of leaves to control larvae.

Lepidoptera
Imported Cabbageworm
Pieris rapae (Linnaeus)

Appearance and Life History

The adult imported cabbageworm, more familiarly known as the white cabbage butterfly, is seen throughout the summer around gardens across North America. The creamy white wings, with black tips on the forewings, have a span of 30-50 mm. Female butterflies have two black spots on the forewings; the males have only one. Eggs are pale white when laid, spindle-shaped and gradually turn yellow. Small velvety green caterpillars hatch from the eggs and grow to be large, 30 mm long, dark velvety green caterpillars, each with a faint lemon-colored stripe down the middle of the back and a slender pale orange stripe along each side. The pupa or chrysalis is smooth, greyish, greenish or tan-colored with sharp, angular projections over the back and front. It is supported on a vertical surface by its tail attached to a silken pad, and by a girdle of fine silken thread around the middle.

Imported cabbageworm larvae feeding on cabbage

M. Herbut

Imported cabbageworms overwinter as chrysalises in crop debris or on buildings or fences. Butterflies emerge about mid-May, mate and lay eggs during the next 2-3 weeks. Each female lays several hundred eggs, one at a time, on the upper surfaces of host plant leaves.

Within a week, larvae hatch and begin feeding on the outer leaves, and mature in 2-3 weeks. When ready to pupate, the caterpillars leave the plants and attach themselves to any vertical surface by means of a single loop of silk and molt to pupae. Adult butterflies emerge in 1-2 weeks during the summer. They can be seen flitting from plant to plant, especially on warm, sunny days when there is little wind. Two to three generations are completed each year.

Food Hosts and Damage

Imported cabbageworms feed not only on cabbage but also on cauliflower, broccoli, Brussels sprouts, canola, and other cultivated and wild crucifers. Other hosts include lettuce, spinach, beets, peas, celery, parsley, potato, tomato, onion and garden flowers such as carnation, nasturtium and mignonette. Larval feeding is characterized by irregular holes chewed in the leaves and in the outer layers of the cabbage head. Partially grown larvae may feed on the inner leaves of the cabbage. They deposit masses of greenish to brown frass throughout the heart of the cabbage head. This feeding and staining of the head with excrement lowers the market value of the crop.

Control

Imported cabbageworms are attacked by at least four species of parasitic insects and a virus disease which has been recorded as suppressing high infestations. The presence or absence of these parasites and/or the disease determines the abundance of the second and third broods. Many larvae are killed when temperatures drop in the fall. One or more larvae per plant requires chemical control. Applications at 2-week intervals may be necessary to produce a marketable crop. A number of products, including the bacterium *Bacillus thuringiensis*, are available to protect plants from feeding damage.

Raspberry Crown Borer

Pennisetia marginata (Harris)

Appearance and Life History

The adult raspberry crown borer is a clear-winged moth which closely resembles yellowjacket wasps. It is black with yellow bands on the abdomen and yellow bands and stripes on the thorax. The forewings are transparent with narrow brown borders and fringes. Females are larger than males, measuring 25-30 mm in length. Eggs are about 1 mm long, oval and reddish-brown. Newly hatched larvae are 3 mm long, whitish with brown heads. Mature larvae are about 25 mm long with a white body and brown head. Each larva has six short legs on the thorax and a series of small, paired, hooked appendages on abdominal segments 3, 4, 5, and 6. These tiny hooks form a pair of oval areas on each of the segments. Pupae are brown and cigar-shaped, about 25 mm in length.

Adults appear in August and September. Within 2 days of emergence, females deposit their complement of 130-150 eggs singly on the undersides of

Raspberry crown borer larva feeding inside raspberry crown

M. Herbut

raspberry leaves, usually two or three per plant. Eggs hatch in September and October. Newly hatched larvae crawl down to the base of the stems or canes where they overwinter inside tiny blisters on the canes just below the soil surface. The following spring the larvae burrow into the crowns where they feed on new and old canes for the duration of the summer. By October the larvae are about 20 mm long and almost full grown. The second winter is spent in their feeding burrows and the next summer they bore further into the fleshy part of the crowns to feed. In July they tunnel a few centimetres up and pupate near the outside of the cane. In August and

September the pupae force themselves part way through the bark of the canes to facilitate adult emergence. Two years are required to complete the life cycle.

Food Hosts and Damage

Raspberry is the primary host of this pest. Larval feeding activity retards cane growth, reduces berry yield, and sometimes kills the plants. During the first summer's development the larvae destroy new shoots. New canes are girdled near the ground resulting in the formation of galls. Canes become spindly and easily broken. During the second summer the larvae continue to feed in the crown and on the roots.

Control

Removing and destroying old and injured canes before mid-August to prevent adults from emerging will help prevent further damage. Smashing old stubs with a mallet will kill any pupae present. A soil drench of a registered insecticide early in the spring and again in the fall will control young larvae. Treatments must be made for at least two successive years.

Insect Pests of Greenhouse and House Plants

Coleoptera
Black Vine Weevil
Otiorhynchus sulcatus (Fabricius)

Appearance and Life History

The black vine weevil was introduced from Europe and now is found across southern Canada. Adults are hard-shelled, roughened beetles about 9 mm long. They are black or brownish in color, usually with small patches of yellow or white scales on the fused forewings. The head projects into a long broad snout typical of all weevils. Larvae live underground and are legless, white grubs, 7-10 mm long, with brown heads. Pupae, which are milky white with conspicuous dusky spines, are also found in the soil and are about the same size as adults.

Adult black vine weevil

M. Herbut

In western Canada there is probably only one generation per year but there may be more under greenhouse conditions. Under outdoor conditions, black vine weevils overwinter as larvae in the soil. After a very active larval feeding period early in the spring, pupation takes place in June, 3-10 cm in the soil. Adults emerge in late June and July and feed on the leaves of trees and shrubs at night. By day, they hide in the soil and in leaf litter under the plants. A 20- to 30-day maturation period takes place while females disperse to new hosts and internal egg development takes place. Adults are flightless and so must walk from plant to plant.

Since males are unknown, females produce eggs without mating. Up to 500 eggs may be laid by each female beneath duff or the soil surface up to 20 cm deep. The number of eggs laid depends on the host plant and the condition of the host on which the adult developed and fed. Egg laying will continue all summer and most females will die before winter. However, some may overwinter and lay eggs early the next spring. Eggs hatch in about 2 weeks and the larvae feed during the summer, fall and again early the following spring.

Food Hosts and Damage

Black vine weevils feed on a wide variety of plants but are pests mainly of evergreen and deciduous plants. Both native and cultivated plants are attacked. They may become pests in nurseries and greenhouses. Rhododendron, strawberry, blueberry, wild and cultivated grapes and cyclamen are listed as some of the more common hosts. Over 100 plants have been reported to support either the adults or the larvae. Adult feeding produces distinctive semicircular notches along the leaf margins and on needles of woody and herbaceous plants, occasionally making them unsightly. Damage by adult feeding is usually far less serious than damage by larvae feeding on roots. Heavy infestations may destroy most of the small feeder roots. As the small roots are destroyed, larger roots are attacked and the crown may be girdled. Destruction of the roots reduces the absorption of water and minerals, causing the foliage to dry out and turn yellow and reduce plant vigor and growth.

Black vine weevils can be a nuisance in containerized and forest seedling nurseries. Greenhouse plants are especially vulnerable. Lifting suspicious plants out of pots and examining their roots and crown for damage is one way

of detecting weevil presence. Tell-tale notches in the leaves and finding the larvae or adults will confirm their presence.

Control

There are presently no good cultural controls for black vine weevils. Placing trap-boards under the most attractive plants and checking for typical notch-type feeding damage of the adults can help in detecting their presence. The use of a residual spray on the foliage of plants at the time of adult wandering before egg laying can reduce problems later on. The insecticide should be applied at night when the adults are active and will kill the beetle by both contact and ingestion. The use of a soil drench on greenhouse material early in the spring before pupation may aid in larval control. This is not effective in nurseries as the chemical must reach 20 cm into the soil and remain undiluted to contact and kill the larvae.

Collembola
Springtails
various species

Appearance and Life History

Springtails are small, wingless insects that live in soil and humus throughout the world. After a heavy rain, puddles of water in farm yards often have a purplish moving scum floating on the surface. If a little of this is picked up on a fingertip, it will be seen to consist of immense numbers of these insects which quickly jump in all directions. Their common name comes from the ability of most species to "spring" or catapult themselves up to 10 cm out of the way when disturbed. They have a fork-like tail, known as a furcula, that is tucked up and held under the abdomen. When released, it strikes against whatever the insect is resting on, propelling the insect through the air and away from danger. It is often the sight of this jumping behavior that is used to identify this insect.

Most species are greyish-blue, with colors ranging from white to orange. Their bodies may be slender or elongate with distinct abdominal segments or globular with fused abdominal and thoracic segments. Adults are usually less than 2 mm in length but some may reach 5 mm in size. They have long, four- to six-segmented antennae but lack the compound eyes of most insects. They do have simple eyes fused to form an eye patch. Chewing mouth parts are located in a pouch on the lower part of the head. A ventral tube is located on the first abdominal segment. The true function of this collophore organ or "glue-peg" is uncertain but may be used in respiration, water resorption or adhesion to the substrate. Respiration is generally through the skin which forces them to live in moist situations.

Adult springtail

M. Herbut

Both male and female springtails occur. Fertilization is external from a spermatophore. Eggs may be laid singly or in clusters under leaf debris. Hatching occurs under cool, moist conditions. Immature springtails look very similar to the adults except paler in color. Springtails do not undergo true metamorphosis so reproductive maturity may be reached after four or five molts with growth and molting continuing after egg laying has begun.

It has been estimated that springtails, in total number of individuals, are the most abundant insects in the world. When their numbers increase, they often "swarm" and are seen on the wall of buildings or on the surface of the snow, hence another common name – "snow fleas".

Food Hosts and Damage

Springtails are found in moist conditions where organic matter is abundant. They feed on live or dead plant material and on molds and algae. Outdoors they occur under rotten logs, boards, wet leaf mold and similar places. They are also found on the surface of stagnant water. Places such as greenhouses, mushroom houses or compost piles are well suited for their development. Springtails can also enter homes in garden soil brought indoors for potting houseplants.

When the environment becomes dry, they begin searching for moisture, entering homes through window screens, open doors or vent pipes. Once indoors, they move about and are often trapped in sinks, wash basins, bathtubs and potted plants. Springtails die soon after entering a home unless they find moisture.

Most springtails are scavengers and are thus beneficial, reducing decaying vegetation and keeping molds and algae down. Some, however, occasionally feed on young, tender plant parts in close contact with the ground, causing numerous irregular holes or pits in the plant. Very large numbers may cause damage to seedlings and potted plants. In houses they are pests principally because of their great numbers but they do not damage household articles. Root cellars and bulb storage areas that are damp may be a cause for concern. Their continued presence in potted soils or households is an indicator of a moist environment.

Control

In homes, springtails are eliminated by airing out and drying infested areas. An electric fan may be helpful and speed up the process. Water leaks or drips should be fixed. Moist leaves and mulch around a building foundation should be eliminated or dried out to prevent invasion from outside.

The use of sterilized soils for houseplants or greenhouse crops may be necessary to prevent springtail introduction and to kill off any springtails present in the soil. Springtail numbers in potted plants or seedbeds may be reduced by allowing the soil to dry out between waterings.

The use of insecticide sprays on or around infested soil or areas may be beneficial if damage is significant.

Diptera
Fungus Gnats
various species

Appearance and Life History

Fungus gnats are small grey-black flies often noticed around house plants and windows. Flies are slender, about 2.5 mm long and have long legs and antennae. Adults may be recognized under a microscope by the eyes which meet over the base of the antennae. Males are slightly smaller than females and the females have an enlarged abdomen when filled with eggs. They are weak fliers but can run rapidly across the soil surface. The tiny eggs are oval, smooth, yellowish-white and about 0.2 mm long (barely visible to the unaided eye). Larvae, or maggots, have shiny black head capsules with chewing mouthparts. The white segmented bodies are 5-6 mm in length at maturity and transparent, showing the digestive systems and its dark contents. Pupae are white at first but gradually become darker. Pupae are 2-2.5 mm in length and may be found in a thin silken cocoon.

Fungus gnat larva

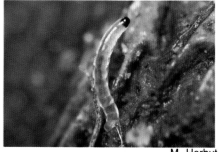

M. Herbut

Adult fungus gnats live less than a week, mating shortly after emergence and laying eggs soon thereafter. Eggs are either scattered or laid in small clusters on the soil surface near plants. Females lay 100-150 eggs in their 2- to 5-day existence. Eggs hatch in 3-5 days and the larvae begin to feed immediately on organic matter near the soil surface. Larvae reach maturity in about 2 weeks when they cease feeding, spin silken cocoons and pupate. The pupal stage lasts less than a week. The entire life cycle only lasts about 4 weeks at greenhouse and house temperatures with many overlapping generations present at the same time.

Food Hosts and Damage

Fungus gnats feed on organic matter, both live and dead. Fungus gnats become a problem when they feed on the roots and root hairs of potted and greenhouse plants, stripping the roots and causing the plant to lose vigor, turn yellow and drop leaves. No visible injury can be seen on above ground parts but roots may have small brown scars on the surface and the maggots may be found feeding within the root system of the plant. African violets, carnations, geraniums, poinsettias and foliage plants are most commonly attacked but all potted plants are susceptible. Fungus gnats are also a problem in mushroom production.

Damage is most severe in plants that have soil that is high in organic matter, especially peat moss. Artificial soil media provide the organic matter that is attractive to adults for egg laying. Swarms of adults become pests by simply their presence as they cause no direct damage.

Control

Fungus gnats prefer moist shady places to live and the reduction of favorable sites will help reduce the population. Allowing potting soil to approach the dry side between watering will prevent a population explosion. An insecticide used as a soil drench and leached through the potting media can be used to control the maggots. Household aerosol insecticides containing pyrethrins will readily kill the adults flying around the house. Regular applications when adults are observed are necessary. Sterilization of new soil will aid in the prevention of introducing the insects into new plants.

Leafminer
Liriomyza trifolii (Burgess)

Appearance and Life History

The leafminer, *Liriomyza trifolii* (Burgess), sometimes called the chrysanthemum leafminer or serpentine leafminer, is a recent introduction to greenhouse crop production in western Canada. Adults are small flies, about 2.5 mm long and dark grey in general color. The underside and legs are pale yellow and a yellow mark is found on the back. The head is yellow with brown eyes. Oval eggs are tiny, 0.1-0.2 mm long, white and translucent. Larvae are white legless maggots, 2 mm in length with a dark head that may be pulled back into the body. The puparia are yellowish-brown and slightly smaller than rice kernels.

Female flies feed by cutting the upper leaf surface with their ovipositor and sucking the sap that oozes out of the leaf. Males also feed at these punctures. After mating, eggs are laid just under the leaf surface in some of the puncture holes. However, only a few of the many punctures or sting holes on a leaf will contain eggs. A female may lay 100 eggs during her 2- to 3-week existence. Eggs hatch in about 4 days and larvae feed on cells just below the upper leaf surface, producing a winding and expanding mine containing black frass. Larval feeding may last 4-10 days, depending on the temperature. When feeding is completed, the larva cuts a hole in the leaf, drops to the soil surface and pupates. Pupation takes about 2 weeks. The life cycle takes 3-5 weeks to complete under greenhouse conditions.

Food Hosts and Damage

The number of plants listed as hosts for this insect is extensive with chrysanthemum, gerbera, tomato and cucumber leading the list of greenhouse crops. Many outdoor crops and weeds are also attacked. Leafminers damage young leaves when females oviposit or sting the leaves, causing white stippling. Destruction of leaves by the mines and retardation of plant growth are the most serious problems. The presence of mines reduces the commercial value of infested ornamental plants and flowers.

Puncture sites may also be entrances for bacterial diseases.

Chrysanthemum leaf damaged by leafminer larvae

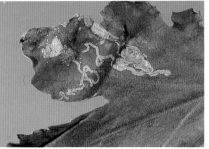

M. Herbut

Control

Cultural controls against leafminers are varied and many. Inspecting new cuttings and transplants and destroying infested leaves before plants are repotted in sterilized soil will prevent the introduction of the pest into the greenhouse. Recognition of damage and alternate hosts within a greenhouse are of utmost importance to prevent its spread. Removal of alternate hosts in and around a greenhouse will help eliminate population reserves. Growing alternate crops not susceptible to or varieties resistant to leafminer may rid a greenhouse of the pest. Yellow sticky boards should be hung in areas where susceptible crops are grown to monitor the insect population and remove some breeding adults. Insecticidal sprays may be used if chemicals that control the fly or larvae are available.

Homoptera
Citrus Mealybug
Planococcus citri (Risso)

Longtailed Mealybug
Pseudococcus longispinus (Targioni-Tozzetti)

Appearance and Life History

Mealybugs are common pests of houseplants and greenhouse crops. Adults are oval-shaped and often are whitish due to a waxy covering on the body. To the naked eye, they appear as though dusted with flour or covered with a cottony, fuzzy material. Adult mealybugs are soft-bodied, wingless insects, about 5 mm in length with well developed legs. There may be two or more long waxy filaments extending from the posterior end of the bodies (as found on the longtailed mealybug). In a few species, winged males may be present. Nymphs are a light yellow color, but as soon as feeding starts, they begin to exude the white waxy material that soon forms a covering over the body. Eggs are laid in masses covered with a cottony material.

Citrus mealybugs feeding on *Coleus*

H. Philip

Adult longtailed mealybug

M. Herbut

Mealybugs are slow-moving, crawling insects, but they will spread readily from one plant to the next. Adults are found at rest or slowly moving on the undersides of the leaves along the veins, on the stems, or in the leaf axils. Some species may be found feeding on the roots below ground. Adults and nymphs feed by inserting their long tube-like mouthparts into plant tissue and sucking juices from the host. The mature female deposits as many as 600 eggs in clusters on the stems and undersides of leaves. Eggs hatch in 1-2 weeks. There are usually three nymphal instars, each lasting 2 or 3 weeks. Males pupate and develop into small two-winged adults that seek out females for mating.

Food Hosts and Damage

Mealybugs have a wide host range and feed on many species of plants. Citrus mealybug not only attacks citrus outdoors in California but also an endless list of plants in the greenhouse. Hosts include especially *Coleus* but also African violets, begonia, geranium, impatiens, *Schefflera* and jade plants. Longtailed mealybugs also attack a similar list of plants but *Dracaena* is the preferred host.

Mealybugs suck out the plant juices, thus stunting or killing the plants. Sooty mold often grows on the honeydew excreted by some species of mealybugs. In addition, citrus mealybug has an apparent toxic effect on the host plant. On *Coleus*, this damage is usually noted when the area along the midrib of the leaf (the area of heaviest infestation) turns brown. Mealybugs may kill a plant but the mere presence of mealybugs gives the plants an unsightly and unhealthy appearance.

Control

Mealybugs can be difficult to control. The waxy cottony masses protect the eggs and the waxy covering protects the young and adults. If only one plant or a small number of plants are infested, hand picking, washing or using a cotton swab dipped in rubbing alcohol will remove individual mealybugs. Heavily infested leaves should be picked off and destroyed. Plants can be dipped in or sprayed with household formulations of suitable insecticides, being sure to wet the mealybugs thoroughly. Care must be taken not to damage foliage on some sensitive plants. Chemical sprays and fumigants for commercial greenhouses are available. Biological control agents such as a predatory beetle (the mealybug destroyer) and a parasitic wasp are other control options.

Greenhouse Whitefly
Trialeurodes vaporariorum (Westwood)

Appearance and Life History

The greenhouse whitefly is not a true fly, but is related to aphids, mealybugs and scales. Adults are small, white insects, 1.5-2.5 mm in length, that resemble tiny moths. As an infested plant is disturbed, a cloud of adults rises from it. Their bodies and wings are covered with a white powdery wax. The four wings are held roof-like over the body when at rest. The oblong eggs are pale green to dark grey in color. Tiny nymphs, yellow with red eyes, become semi-transparent scale-like insects flattened against the lower leaf surface. "Pupae" are flattened, oval and pale green with a fringe of long hairs and with long hairs on their backs. All feeding stages have sucking mouthparts.

Greenhouse whitefly adults and pupae

M. Herbut

Greenhouse whiteflies reproduce slowly with one generation every 30-45 days. Each female lays 250-400 eggs at a rate of 25 per day. She may live 1 or 2 months. Eggs are laid in the leaf tissue in a circle or crescent on the undersurface of new leaves. Eggs hatch in 5-10 days. The first stage nymphs, or crawlers, move about a plant for a day or two before inserting their mouthparts to feed. Once feeding begins, they molt into legless nymphs that resemble young scales and remain in one spot on the leaf. The "scale" thickens after feeding for some time and passes through a quiescent stage, often incorrectly referred to as a "pupa". Winged adults emerge in 4-5 days. Several overlapping generations occur each year on greenhouse crops with all stages present at any time. Greenhouse whiteflies do not survive outdoors during the prairie winters but may move outside during the summer.

Food Hosts and Damage

Greenhouse whiteflies infest a wide variety of ornamental and vegetable crops grown under glass and in the house. Some of the more important crops include tomato, cucumber, lettuce, poinsettia, fuchsia, geranium, gerbera and chrysanthemum. Severe feeding by these tropical sucking insects causes stunting, distorted leaves and weak plants. Whiteflies secrete a sticky substance, called honeydew, that makes plants unsightly when in excess, and also serves as a development site for a black, sooty mold. The mold also reduces the amount of photosynthesis taking place in the leaf thus reducing plant vigor.

Control

Inspection of new plants and isolation of infested plants may aid in preventing the introduction of whiteflies into a new area. Adults cannot survive more than a week without food plants and therefore a break between spring and fall crops and clearing the greenhouse of all plants, including weeds, will greatly reduce or eliminate whiteflies. The parasitic wasp *Encarsia formosa* has been used successfully as a biological control agent in greenhouses and indoor plant scapes.

Chemical control is difficult because of the complex life history of the greenhouse whitefly. The five distinct developmental stages have different tolerances to insecticides. Eggs are resistant to all but a few insecticides. Crawlers are susceptible to insecticides. The "pupae" are non-feeding and immobile, and not susceptible to insecticides. Since adults fly from plant to plant seeking new sites to lay eggs, they may be controlled by fumigants, space sprays and contact insecticides. A single application of a particular insecticide only affects susceptible stages and leaves the other stages to survive and reproduce. Several treatments at 5- to 7-day intervals are required to control all life stages. Sprays must be directed to the undersides of leaves. Treatments must begin as soon as whiteflies are noticed.

Scale Insects
various species

Appearance and Life History

Scale insects derive their name from the scale or shell-like appearance of adult females. They average about 3 mm or less in length, oval to elongate in shape, and vary in color from white to dark brown. Scales are divided into two general groups: armored and unarmored scales (also called soft scales). Female armored scales are wingless, legless, eyeless insects covered with a convex shell of secretory material or wax under which they live and feed. Female unarmored scales are also immobile, but do retain legs throughout their lifetime, and the scale is not made of secretory material but of hard body wall or cuticle, and therefore is not separable from the body and is leatherlike. Adult male scales develop under thin shells and emerge as delicate, usually two-winged insects without functional mouthparts. After mating with females, they disappear, leaving the females to lay eggs or give birth to live, mobile young (crawlers) under their shells or scales. It is during this early stage that the crawlers of most scale insect species disperse on and among suitable plant hosts. They may walk about on the plant, be blown by the wind, or be accidentally carried about by man, animals, or other insects. When ready to feed, the nymphs molt and the shell or waxy covering is formed. Once having settled down to feed, the crawlers of most scale species lose the use of their legs and remain at the same position on the plant for the remainder of their life span.

Outdoors, scale insects require 2 months to develop from egg to adult stage with only one generation being produced annually. In greenhouses and in households, several generations can be produced in a year.

Food Hosts and Damage

Scale insects are a more serious pest of greenhouse and house plants than outdoor plants. Some species are specific plant feeders, whereas other species feed on a variety of plants. Scale insects appear as small, off-colored spots on infested plants, often causing discoloration and wilting.

Scale insects feed by sucking sap from plants and are capable of killing the entire plant or parts of the plant. Only females cause damage as adult males do not feed. Scale insect feeding can also reduce the plant's vigor, which will make the infested plant more susceptible to injury caused by drought, severe winters, or attack by other insects or infection by diseases. Some species insert a toxin into the plant with the saliva which may cause damage to the plant. Soft scales may produce a sticky secretion known as honeydew that covers the leaves and may attract fungi and other insects such as ants to the plant.

Control

Natural enemies play a major role in controlling scale insect populations and preventing scale insects from causing significant plant damage. Natural parasites have been used extensively outdoors and are becoming available for greenhouse use also. Cultural controls may include pruning heavily infested branches before crawlers hatch, misting or spraying with water to wash off honeydew and dislodge crawlers, isolation of infested plants in the greenhouse, and hand crushing the scales when first seen. Dormant oils may be applied before bud break in the spring to control ornamental scales by suffocation. Insecticidal sprays may be applied to the crawler stage as most sprays will not penetrate the waxy scale covering. Several applications may be required.

The pine needle scale, *Phenacaspis (Chionaspis) pinifoliae* (Fitch), is a pest mainly of white and Colorado spruce and Scots pine in nurseries, shelter-belts and ornamental plantings. The scales are white, elongate to pear-shaped, about 3 mm long, and located along the needles. Heavy infestations cause needles to turn yellow, and can cause death of individual branches. They overwinter as eggs beneath the scale, hatching in the spring and dispersing to other needles and hosts. Mating and egg laying occur in August; there is only one generation per year. Warm, dry conditions favor pine needle scale development and damage. Chemical controls may be applied in the spring when the crawlers are dispersing and again in early August before females form the hard shell.

White spruce needles infested with pine needle scale

M. Herbut

The European fruit lecanium, *Lecanium corni* Bouché, appears as a pale to dark brown, spherical lump on the small branches and twigs of flowering plum, elm, variegated dogwood, false spirea, white birch, ash, some apple varieties and many other trees and shrubs. Females mature in early summer, mate and produce eggs. The eggs remain under the scale from mid-June until late July when they hatch into yellowish-brown, flattened crawlers which immediately move to the leaves of the tree where they feed until fall. They settle along the leaf veins and secrete a scale over themselves. However, before the leaves drop in the fall, they crawl off the leaves and attach themselves to the twigs where they overwinter. A heavy infestation will cause the leaves to become discolored and reduce the tree's vigor, possibly leading to the death of small branches and twigs. Controls are only necessary if the tree or shrub is suffering severe damage.

Green ash infested with European fruit lecanium

M. Herbut

The oystershell scale, *Lepidosaphes ulmi* (Linnaeus), resembles a tiny oystershell on many deciduous shrubs and trees, especially lilac and cotoneaster. The grey or dark brown scales may become so numerous that stems are encrusted, causing branches to become spindly and leaves to turn yellow. Oystershell scales overwinter as eggs under the protective covering of the dead female scales. Eggs hatch in spring, the crawlers move to nearby stems, insert their mouthparts and feed. The waxy protective coat is secreted and the insect becomes fully grown by mid-July. The female then begins to lay eggs under the scale.

Cotoneaster twig infested with oystershell scale

M. Herbut

The scurfy scale, *Chionaspis furfura* (Fitch), forms a small, dirty white, pear-shaped scale about 2 mm long on the leaves, branches and trunks on a wide variety of trees including elm, willow, ash, maple and apple. Dormant oils or summer sprays applied in early June to control the crawlers may be useful.

Crabapple twig infested with scurfy scale

H. Philip

The Boisduval scale, *Diaspis boisduvalii* Signoret, lives on the leaves and bark of orchids, palms, bananas and cacti in greenhouses. The shield of this armored scale is circular, white and flattened, with a central nipple. The males are smaller and elongate with three powdery ridges along the back and they cluster together in cottony masses. Pruning infested branches and misting to discourage crawlers aid in eliminating or retarding population build-up. Biological and chemical controls are available for this insect.

Boisduval scale feeding on orchid

M. Herbut

The hemispherical scale, *Saissetia coffeae* (Walker), is a smooth, strongly convex, shiny brown scale, roughly circular in outline. Immature stages are white with an "H" pattern on the upper surface. The underside of the female is cupped to accommodate large numbers of eggs. They attack many plants including fern, *Schefflera* and asparagus fern. This soft scale, common in greenhouses on perennial crops, can be controlled similarly to the Boisduval scale.

Female hemispherical scale with immature scale and crawlers

M. Herbut

The brown soft scale, *Coccus hesperidum* Linnaeus, is oval and more flattened than the hemispherical scale. The soft shell is pliable, pale brown, dirty white or greyish, and mottled with dark brown on the back. Brown soft scales feed only on phloem tissue, doing very little damage to the plant but they do produce abundant honeydew which promotes the growth of sooty mold. Females produce living young rather than eggs and so crawlers may be present at all times in the greenhouse. Reproduction is continuous and rapid under greenhouse conditions; six or more generations are produced each year. Brown soft scales feed on many plants including *Schefflera*, weeping fig, citrus and ferns.

Female and immature brown soft scale

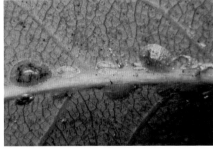

M. Herbut

Thysanoptera
Thrips
various species

Appearance and Life History

Thrips are very small, slender insects with most species only 1-2 mm long. Adults are flattened insects with four long, strap-like wings fringed with long hairs. Colors vary from yellow and orange to tan, dark brown and black. Eggs are bean-shaped and large in relation to the size of the insect. Young thrips resemble adults except for their smaller size, lack of wings and paler color.

Most important species of thrips overwinter as adults in plant debris in fields, along headlands or under bark. Adults become active early in the spring and lay their eggs in slits which they cut in leaves or flower petals of suitable host plants. Eggs hatch in about 7 days and the minute larvae remain actively feeding on the plants. The development of thrips is intermediate between gradual and complete metamorphosis. The larvae pass through four instars or stages, the first two are actively feeding and the last two, the prepupa and pupa, are relatively inactive and non-feeding. Most species pupate in the soil, however, a few pupate freely on leaves or in silken cocoons attached to leaves. After a few days, the adults emerge, mate (if males are present), and either remain on the original host plant or fly to a more suitable host plant to start another generation. Several generations are produced annually, each one requiring 3-4 weeks to complete, depending upon species and temperature.

Food Hosts and Damage

Due to the small size of these insects, their presence is most easily recognized by the damage they do to host plants. They feed with "punch and suck" mouthparts on the outer plant tissue. They first punch a hole in the plant cell and then suck up the juices. The damage appears as whitish blotches or streaks on leaves, flower buds and petals. In crops grown for seed, the flowers may be damaged enough to prevent seed production, thus reducing yields. A few species of thrips are predacious and attack other thrips, small insects and mites.

During their flight to other host plants, thrips may appear in swarms and have been reported as biting people. They are not seeking blood but rather water, and the bite is often itchy but generally harmless.

Control

Thrips like moist, humid environments and with their small size, they can easily hide in leaf sheaths and unopened buds. This makes them especially difficult to control with insecticides. Since thrips can occur in a number of different habitats and on a number of different kinds of crops, controls and timing must be directed to that particular thrips.

The western flower thrips, *Frankliniella occidentalis* (Pergande), is primarily a pest in greenhouses. Adults are orange-brown to straw-colored and 1-2 mm long. They normally feed on the underside of leaves and lay eggs in the foliage. Feeding on the developing fruit of cucumbers causes scarring which results in crooked and unmarketable fruit. After maturing, the larvae drop to the ground to pupate. The life cycle (egg to egg) may take from a month at 18°C to 2 weeks at 30°C. Several applications of insecticides are required to control the active feeding stages (adults and young larvae). Soil-applied insecticides to control pupal stages and emerging adults are of some benefit.

Adult western flower thrips

M. Herbut

The onion thrips, *Thrips tabaci* Lindeman, occurs across the southern prairies and has been recorded as a pest of asparagus, bean, cucumber, cabbage, cauliflower, turnip, green pepper, carrot, potato, corn, alfalfa, clover and particularly onion. Blotches of white or bronze and roughened texture on developing leaves are signs of thrips presence. Small black spots of fecal droplets are also evidence of thrips. Damage is greater during warm dry weather.

Onion thrips adults (brown), pupa (white) and larva (yellow)

M. Herbut

The grass thrips, *Anaphothrips obscurus* (Müller), can reach very high populations on late-seeded barley, other cereals and grasses. They feed within the sheath and panicle, and are one of the causes of a condition known as "silvertop".

The red clover thrips, *Haplothrips leucanthemi* Schrank, feeds in the flowers of red and sweet clover and alfalfa, and can significantly reduce seed yields when numerous in these crops.

Greenhouse thrips (various species) attack and damage plants grown indoors, especially greenhouses. These thrips cause damage by feeding on the flowers producing streaks and browning on petals, and by feeding on the foliage. The presence of tiny black fecal drops on plants is also evidence of their presence. Some of these species pupate on the leaves and therefore the application of an insecticide to the soil surface is ineffective. Detecting and controlling thrips before plants flower is important to prevent blossom damage.

Insect Pests of Ornamental Trees, Shrubs and Turf

Acari
Mites
various species

Appearance and Life History

Mites are very minute animals, rarely more than 1 mm long, and closely related to insects. Their small size makes them difficult to see with the naked eye. They differ from insects in having only two body sections and usually four pairs of legs. They lack wings and so must either walk, be blown in the wind, or be carried to new host material. Their life cycle includes egg, larva, nymph and adult stages.

Depending on the species, mites overwinter either in the egg, nymphal, or adult stages in protected situations such as under bark or debris. They become active in spring as new foliage begins to appear on host plants. After mating, female mites lay their minute spherical eggs on the host. Six-legged larvae hatch from the eggs. After feeding for awhile, the larvae molt into an eight-legged nymph. The nymphs molt two or three times before the adult stage is reached. There may be several overlapping generations per year outdoors. Indoors, reproduction takes place year round, so mites may be a continuous problem on house plants and greenhouse plants. Plant-infesting mites often feed on the lower surface of the leaves. Their rasping-sucking mouthparts penetrate leaf surfaces and remove cell contents, causing a type of injury called "stippling".

Food Hosts, Damage and Control

The twospotted spider mite, *Tetranychus urticae* Koch, often called the red spider mite or glasshouse spider mite, is the most commonly encountered mite causing damage to plants. Less than 0.5 mm long, it is difficult to spot without the aid of a hand lens. The body is oval and yellowish or greenish except for two dark spots on the back. Outdoors, females hibernate in leaf litter or sheltered situations such as under tree bark. They emerge in late spring and each female lays about 100 circular eggs during the first month, usually on the lower leaf surface along the midrib. Summer generations of mites may live about a month but under optimum temperature (30°C), a new generation is produced every 8 days. At minimum temperature (12°C), development may take up to 6 weeks. Thus, hot dry weather favors a very rapid increase in populations. As the temperature drops in late summer, overwintering females develop and seek shelter for the winter.

Twospotted spider mites attack a wide range of indoor and outdoor plants. Most greenhouse plants, many ornamental trees, shrubs such as roses and alder, berry fruits such as raspberries and strawberries, and vegetables such as cucurbits and legumes are susceptible. Mites pierce the epidermis and remove the cell contents, causing the leaves to become speckled or spotted. Leaves eventually turn dull to pale yellow and even bronzed or reddish with very high mite populations. Leaves may dry up and drop from the plant. This results in reduced yield of the crop and unsightly plants. Webbing may be found on heavily infested leaves and stems.

Control of twospotted spider mites may include misting of plants to increase humidity, removal of favored plants to reduce infestations, removal of host weeds in greenhouses, lowering the temperature to slow down development, and inspecting plants regularly to find new infestations before they build. Predatory mites may be used under some situations. The use of chemical miticides at regular intervals may also be useful.

Adult and immature twospotted spider mites with eggs

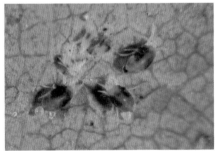

M. Herbut

The cyclamen mite, *Steneotarsonemus pallidus* Banks, is a pest of such plants as cyclamen, African violets, gerberas, ivy and strawberries. This tiny mite, less than 0.3 mm long, is yellowish-brown. Females may lay anywhere from 20 to 100 eggs in dark, moist places in the buds and leaf clusters of host plants. Several generations may be produced during the summer. Damage is characterized by twisted and stunted leaves and flowers from mites feeding in the developing buds. Controls are especially difficult because the mites hide in the closed leaf and flower clusters and penetration by a miticide into these areas is almost impossible.

African violet plant damaged by cyclamen mites

M. Herbut

The poplar budgall mite, *Eriophyes parapopuli* Keifer, is a very tiny elongate mite that attacks poplars on the prairies. During feeding, a substance is injected into the bud, stimulating the formation of a woody gall, stopping twig growth and reducing foliage development. The mites live inside the gall, feeding on gall tissue and producing a new generation every 2-3 weeks during the summer. Galls continue to enlarge for several years until the gall may be 2-4 cm in diameter. Mites overwinter in the gall and move to new buds in the spring, feeding on the leaves and causing a cauliflower-like swelling that will turn red in late summer. Older galls are grey, hard and woody with deep folds and ridges. Pruning galls as they develop, especially on small trees, will remove the overwintering mites and prevent their spread.

Poplar budgall mite damage to poplar

M. Herbut

The spruce spider mite, *Oligonychus ununguis* (Jacobi), can be a serious pest of conifers such as white and Colorado spruce, fir and junipers. This species overwinters as eggs near the base of needles or under bud scales. Eggs hatch in mid-May with larvae and nymphs feeding on the needles. Several generations are produced each summer at 2- to 3-week intervals. Females lay

40-50 eggs during their lifetime. Spruce spider mites feed on the needles, causing them to turn yellow, then brown and finally drop off. Feeding damage is greatest on the lower branches and begins near the centre of the tree, progressing outwards along the infested branches. Mites also spin a fine webbing between the needles that traps dirt and dust particles giving the tree an unhealthy appearance. Spruce spider mites are usually controlled by rain and wind or by washing the trees regularly with a high pressure stream of water. They may be controlled by one or two well-timed applications of a miticide directed at the lower branches and inner needles.

White spruce twig infested with spruce spider mites

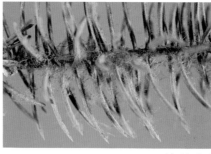

M. Herbut

White spruce tree damaged by spruce spider mites

H. Philip

Eriophyid mites (various species) are also known as blister, gall or rust mites. These elongate white mites are extremely small (0.1-0.3 mm in length), with two pairs of legs at the front end. Most are very host specific, attacking only a few tree species and producing very distinctive galls. They do not usually cause any significant damage to trees but the local abundance and often bright colors of their feeding damage make them very noticeable. Eriophyid

mites on deciduous trees usually overwinter as adults under bark scales. In the spring adults move to developing leaves and begin feeding. Feeding activity stimulates cell division and gall formation by the leaf. Bladder and spindle galls enclose the mites while colorful rust galls provide a rough surface for the mites to feed on. Mature mites may leave the gall and crawl or be blown in the wind to new leaves to start feeding again. Once a gall has been formed, the damage to the leaf is permanent until the leaf falls. Controls are rarely necessary and usually difficult because the galls protect the mites from any pesticide that may be sprayed on the plant. Application of a miticide or a dormant oil early in the spring before the galls are formed may be helpful in reducing the damage later in the summer. A systemic pesticide may be useful in killing mites in previously formed galls.

Eriophyid mite galls on American elm leaves

H. Philip

Birch leaves damaged by eriophyid mites

M. Herbut

The ash flower gall mite, *Aceria fraxiniflora* (Felt), attacks the male flowers of green ash trees, causing them to develop abnormally and form irregular galls. The flowers turn black and remain on the tree after the leaves are shed, making them especially conspicuous on affected trees during the winter. Damage is sporadic because ash does not produce flowers every year. Mites overwinter as females under bark and in cracks throughout the tree. In the spring, they move to the developing flower buds, where eggs are laid. Several generations are produced each year. Controls are not usually necessary. However, spraying may be required on high-value trees.

Coleoptera
Ash Bark Beetles

Hylesinus spp.

Appearance and Life History

Three species of ash bark beetles cause damage to ash trees on the prairies. The grey and brown adult beetles are 2-3 mm long. The larvae are C-shaped, legless and white with a brown head.

All 3 species have similar life cycles. They overwinter as adults either in hibernation chambers (hibernacula) in the bark at the base of the tree or in tunnels between the bark and wood of infested branches. In the spring, females construct egg galleries or tunnels perpendicular to the branch or trunk and between the bark and wood of dying or severely weakened trees. The western ash bark beetle, *Hylesinus californicus* (Swaine), chews ventilation holes through the bark at frequent intervals. Eggs are laid in niches along each side of the galleries. The larvae tunnel into the inner bark of the trunk and branches, parallel to the grain of the wood. Adults emerge in late summer and form overwintering hibernacula. There is normally only one generation per year.

Food Hosts and Damage

Green, black and Manchurian ash (*Fraxinus* spp.) are all attacked by ash bark beetles. The first sign of beetle attack is the "flagging" in the upper crowns in late July and early August. Beetle burrowing beneath the bark girdles the branches and trunks, causing the leaves to turn yellow, red-brown and then die. As the larvae extend their tunnels, the bark becomes sunken and discolored and marked by encircling

rows of holes about 1 mm in diameter and about 1-4 mm apart. These ventilation holes are often joined by a fine, hairline slit. In the year following the attack, the bark between the holes deteriorates, and eventually collapses, revealing the gallery beneath. As an attack advances, branch dieback continues and the beetles may move from branches into the trunk, completely girdling the tree and causing its death.

Green ash branch showing damage (holes) by the western ash bark beetle, *Hylesinus californicus*

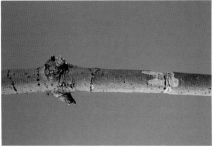

M. Herbut

Control

Summer pruning of infested branches showing entrance and ventilation holes may aid in reducing future damage. Prunings should be burned to destroy any larvae or beetles remaining in the branches.

Bronze Birch Borer

Agrilus anxius Gory

Appearance and Life History

The bronze birch borer is a native insect that may be causing more damage to birch trees in urban areas than is realized. Adults are slender, 7.5-11.5 mm long, olive- to copper-bronze-colored beetles. The greenish-faced males are slightly smaller than the copper-faced females. The flattened, creamy white oval eggs measure 1.5 mm long by 1 mm wide. Larvae are white, narrow, flattened and legless. The light brown head capsule is retracted into the slightly widened thorax. Two brown, hardened spines are located at the end of the abdomen. Mature larvae measure up to 35 mm in length. Pupae resemble adults but are creamy white at first and gradually darken to a brownish-black.

Bronze birch borer larva tunneling in birch

Northern Forestry Centre, Canadian Forestry Service, Edmonton

Eggs are laid singly or in small clusters in the bark crevices or beneath loose bark flakes on the trunk or branches of host trees in mid-summer. Most eggs are laid on the sunny side of the tree but may also be laid where recent mechanical or other injury makes a more attractive egg-laying site. Eggs hatch in about 2 weeks and the young larvae bore into the tree to feed in the inner bark (phloem) and cambium, creating meandering tunnels between the wood and bark and occasionally forming small cavities in the wood to molt or overwinter. The tunnels are packed with sawdust and excrement which turns dark brown with age. Larvae may feed 1 or 2 years before building oblong cells just beneath the cambium in which to pupate. Pupation takes place in spring and adults emerge from June to August. Emergence from the bark is through distinct 4- to 6-mm wide semi-circular or D-shaped holes. Adults feed on the foliage for a week or more before mating and egg laying. One or 2 years may be required to complete one generation.

Food Hosts and Damage

European white birch and its cutleaf varieties are extremely susceptible to damage with all birch species, both native and introduced, serving as hosts. Tree injury is caused by excessive larval tunneling under the bark and cambium (the region of tree diameter growth) and the girdling of the trunk or branches interrupts sap flow downward to the roots and destroys the tree's cambium tissue. The interruption and subsequent accumulation of sap flow above larval tunnels often causes characteristic swollen bands or ridges in trunks and affected branches. The first sign of borer attack is a die-back of the uppermost branches followed by a gradual decline and eventual death of the entire tree in 2 or 3 years. The presence of the D-shaped adult emergence holes in tree trunks is a sure sign of attack. The removal of bark from infested trees reveals irregular, winding, sawdust packed tunnels. Foliar damage by feeding adults is not considered serious and does not affect the condition of the tree.

Control

Maintaining strong healthy trees by proper watering and fertilizer application is the best line of defense against the bronze birch borer. Natural predators and parasites seem to be ineffective in the urban setting. Control of birch leafminer and aphids will assist in maintaining tree vigor but the effectiveness of soil drenches of systemic insecticides against bronze birch borer is not certain. Dead and dying branches and tops must be pruned below the damaged area and the prunings destroyed to prevent reinfestation from woodpiles. Trees with one-half the crown destroyed should be cut and replaced with less susceptible trees. Care must be taken to avoid mechanical damage to the trees that will predispose the trees to attack.

Cottonwood Leaf Beetle
Chrysomela scripta Fabricius

Willow Leaf Beetle
Calligrapha multipunctata multipunctata (Say)

Appearance and Life History

The cottonwood leaf beetle and willow leaf beetle are only two of several leaf beetle species that feed on the leaves of broadleaved trees across the prairies. Adult cottonwood leaf beetles are 8 mm long, slightly oval in shape and only moderately convex. They have black heads with a black thorax edged in yellow or red, and have yellow wing covers with seven elongated dark spots. Eggs are bright yellow. Mature larvae are blackish and may reach 12 mm in length.

There are two whitish spots on each side of each segment at the site of the scent glands that emit a pungent odor when the larvae are disturbed. Pupae hang head down from leaves, bark, or weeds and grasses under the host trees.

Adult cottonwood leaf beetle feeding on poplar leaf

M. Herbut

Cottonwood leaf beetle larvae feeding on poplar leaf

M. Herbut

Willow leaf beetle adults are pale yellow to white with many irregularly arranged dark spots on each wing cover. The head and back part of the pronotum are black. Beetles are 6-8 mm long. Full-grown larvae are about 10 mm long, with black head and legs. The body is white with some small black spots and the abdomen is conspicuously swollen at the middle. Pupae are found in the soil.

Adult willow leaf beetle

M. Herbut

The life cycles of these two species are very similar. They overwinter as adults under bark, litter, and forest debris. Adults become active in spring, and begin feeding on unfolding leaves and on tender bark of host trees. Females deposit their eggs in clusters on the undersides of leaves. Young larvae feed in groups whereas older larvae feed singly. Larvae feed for 2-3 weeks, then pupate. Adults emerge in about 2 weeks. One or more generations may occur on the prairies each year.

Food Hosts and Damage

Cottownwood and willow leaf beetle adults and larvae feed on poplar and willow. Feeding damage is characterized by skeletonized and stripped leaves. Cottonwood leaf beetles feed primarily on cottonwood poplar and will not feed on aspen poplar. They will also destroy the apical growing tips causing stunted growth and deformed trees. Willow and alder may also be attacked. Both insects can be destructive in nurseries.

Control

Natural enemies such as lady bird beetles feed on the eggs and young larvae. Cold winters with little or no snow cover cause very high adult mortality. Chemical control is not usually necessary except in tree nurseries, or when ornamental trees are being severely defoliated. Larvae can be washed off small trees using a high-pressure water hose.

June Beetles
Phyllophaga spp.

Appearance and Life History

June beetles are large (20-25 mm long), robust, brown or brownish-black beetles with prominent club-shaped antennae. The pearly white eggs are oblong, 2.5 mm long by 1.5 mm wide. Larvae, commonly called white grubs, are white and always bent in a crescent or C-shape with the fleshy legless abdomen lying under the hard brown head. Mature white grubs reach a length of 25-30 mm. Pupae are oval-shaped and brownish in color.

June beetle larva or white grub

H. Philip

Adult June beetles emerge in early May from the soil where they overwintered. They are active only at night, flying in search of suitable trees on which to feed. Through late May and June, females lay their eggs at depths of 3-10 mm in the soil. The eggs hatch in 2-3 weeks and the young white grubs feed on the roots and underground parts of plants until early fall when they are about 12-13 mm long. The grubs pass the winter in the soil just below the frost line.

The following spring they move upward as the soil warms up, and again feed on roots and underground stems. In early fall of the second year they again burrow into the soil to below the frost line to overwinter. In the third summer the grubs pupate in earthen cells in the soil. The subsequent adults do not leave the soil immediately but remain inactive 15-20 mm below the soil surface until the following May. Life cycles of the most destructive species of June beetles take 3 years to complete.

Food Hosts and Damage

White grubs live naturally in grasslands and are usually most plentiful in light and acid soils. They become an economic problem in crops sown on newly broken grassland, feeding on seedlings and transplants of such plants as timothy, wheat, corn, potato and most vegetables. Plants attacked to a lesser degree are barley, oats, rye, orchard grass, peas, beans, canola, buckwheat and some clovers. Twenty or more larvae per square metre are considered potentially dangerous to crop yields.

Potatoes are very susceptible to white grub infestations. The larvae chew shallow grooves and cavities in the tubers thereby increasing the chances of fungal or bacterial disease infection and reducing marketable yields. On non-tuber crops, grubs feed on fine root hairs and roots. Damage to grain is characterized by stunted or dwarfed plants and reduced yields.

White grub feeding damage to lawns is noticeable during dry weather as wilted and brown patches of grass. Damaged turf may die out in patches and become so loose that it can be rolled up like a carpet. Garden and ornamental plants may become unthrifty or die as a result of white grub feeding injury.

Adults eat the leaves of deciduous trees and shrubs, and are a nuisance in buildings and around lights during their flight periods.

Control

Grubs are attacked by a wide range of bird, insect, nematode and fungal predators as well as skunks and moles. Maintaining a good stand of turf grass is an important preventive measure. Avoid planting susceptible crops in recently plowed sod. Chemical controls are seldom necessary except for severe infestations. Surface-applied insecticides followed by irrigation will move the insecticide into the soil where the grubs live.

Mountain Pine Beetle
Dendroctonus ponderosae Hopkins

Appearance and Life History

The mountain pine beetle is a native insect of the lodgepole pine forests of western North America. Adults are small, 5-7 mm long, stout, cylindrical black beetles. Eggs are ovoid, white to cream-colored and less than 1 mm long. The four larval instars are white to cream-colored with amber head capsules. Pupae are white to cream-colored and of the general form and size of the adult. Legs and wing pads are folded beneath the body, and the abdominal segments are exposed.

There is usually one generation per year. All stages are spent under the bark of infested trees except for a few days when adult beetles emerge and fly to attack new trees in late July and August. Adults select and attack living trees, boring through the bark and constructing vertical egg galleries. Females excavate narrow galleries 15-70 cm long on the stem, along which she lays 60-80 eggs in niches arranged singly in alternate groups along the sides of the gallery. Eggs hatch in 1-2 weeks and the larvae feed in the phloem, usually making tunnels at right angles to the egg gallery. The larvae develop part way before winter, and then resume development in the following spring. By late June they have grown to full size, 6-8 mm long, and prepare small oval excavations in the bark in which they pupate. Here they transform to pupae and then to adults, thus completing the life cycle. The young adults chew circular holes through the bark to exit the tree, and then take flight to seek new host trees. Some variations to this occur at high elevations and early flying females may actually make two galleries and some may overwinter.

Food Hosts and Damage

The main hosts are the ponderosa, lodgepole and white pine with other pines of the forested regions being attacked occasionally. Mountain pine beetle has attacked lodgepole and Scots pine in urban and park areas across the southern Canadian prairies and may attack jack and mugho pine. Exotic pines seem to be more attractive and vulnerable than native pines. Usually the beetles attack trees older than 60-80 years and with stems over 20 cm in diameter. However, when abundant or when only smaller pine trees are available, the beetles may attack and kill younger trees with stems only 10 cm in diameter.

The first signs of infestation are the pitch tubes on the trunks of living trees. These pitch tubes appear during the summer months, and mark the places where female beetles have entered the tree. Pitch tubes are cream-colored to dark red masses of resin, mixed with boring dust, and are 6-12 mm in diameter. The pitch tubes are the tree's defense against attack and an attempt by the tree to repel the insect from the tree. Dry boring dust similar to sawdust in bark crevices and around the tree base indicate a successful attack by the beetle. The pitch tubes also mark the lower end of the adult gallery.

Depending upon the density of beetle attacks per tree stem, the vigor of the tree and its relative resistance to attack, death of the tree usually follows within a year. Death results from the girdling effect of the larvae as they feed in the bark and from a specialized group of fungi which develop in the bark and sapwood of the tree stem. The spores of this fungal complex are always associated with mountain pine beetles and are carried by the beetles when they bore into the tree. With colonization of the fungi in the sapwood, the translocation system of the tree becomes blocked and the sapwood takes on a characteristic bluish color as the tree dies.

Foliage of an infested tree begins to fade to a yellowish color within 10 months after attack (May and June of the following year), then to a rust-brown or straw color. The tree is essentially dead at this point but may still contain beetles under the bark. Trees which are resistant in one year may be reattacked the following year, at which time they may or may not succumb.

Lodgepole pine trees (brown) killed by mountain pine beetle

Northern Forestry Centre, Canadian Forestry Service, Edmonton

Control

Control is a difficult and costly process and not always successful. Early detection of the beetles and removal of infested trees before the beetles emerge is the most widely practiced method of control. Trees which have become infested in the previous year and have immature stages (eggs, larvae, pupae)

present under the bark and fading foliage should be removed and destroyed before the beetles emerge in July or August. In most cases such trees may already be dead and application of an insecticide at this stage would be of little help. However, where individual trees have a high value such as in parks, shelterbelts, campgrounds and as ornamental plantings, application of a chemical insecticide can provide some protection. Water-based insecticidal sprays applied to the lower half of the tree trunk in mid-July can provide season-long protection from the adults.

Poplar Borer
Saperda calcarata Say

Appearance and Life History

The poplar borer is the most destructive wood borer of poplars. Adults are large (20-30 mm long), sturdy beetles, clothed with dense grey hairs. Their very long antennae and orange-yellow markings on the wing covers are useful for identifying this insect. Eggs are creamy white, oval in shape and about 4 mm in length. Mature larvae are large (50 mm long), dirty white legless grubs which feed under the bark of poplar trees. Pupae are yellowish-white, 20-35 mm long, and found in the boring galleries.

Poplar borer larva

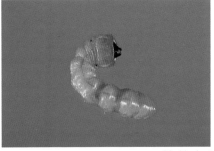

M. Herbut

Adults emerge in late June and July and feed on poplar and willow leaves, petioles and bark of tender twigs. Mating occurs about 1 week after emergence and egg laying begins about 1 week later. Females gnaw crescent-shaped niches in smooth stem bark at a slight angle to the main stem axis of the tree. Eggs are laid singly in the bark and then sealed into the bark with a secretion. Eggs hatch in about 2 weeks. The

first-instar larvae feed on inner bark and sapwood, forming a small chamber in which to overwinter. Larvae resume feeding in early May, extending the tunnel inward to the heartwood. Frass and wood fibers are ejected from the tunnel through enlarged oviposition holes or through new holes chewed through the bark. Feeding in the sapwood may continue for 1 or 2 years. When mature, a fibrous frass plug is inserted into the opening of the tunnel. The larvae construct pupal cells near the lower end of the larval mines and remain inactive until the following spring. Pupation occurs in the larval tunnels and the adults emerge through the holes used by the larvae for expelling boring dust.

Food Hosts and Damage

Poplar borers attack the trunk and branches of living tree hosts (trembling aspen, plains cottonwood and balsam poplar) and are found wherever these trees are present. Females tend to lay their eggs in smaller aspen trees (4.5-18 cm in diameter) and these may occasionally be killed by the girdling of the larvae under the bark. However, little direct tree mortality occurs from poplar borer damage and feeding. Greater damage results from the decay that becomes established in the abandoned mines and the breaking off of trees where the heartwood has been weakened. Poplar borer damage can be recognized by the presence of swollen scars, holes in the trunk and larger branches of the tree with sap and wet areas around the holes giving a blackish or varnished appearance to the bark. Piles of sawdust and chewing dust are also found around the base of the tree. Woodpeckers may also begin attacking the trees.

On balsam poplar, the attack is near the base of the trunk at ground level. A definite gnarled bulbous swelling of the bark may be observed on the stem above and adjoining roots below ground level. The presence of reddish-brown fibrous material at the base of the stem is also indicative of attack.

Poplar borers tend to attack faster-growing trees in an area and certain trees, known as brood trees, are more attractive than others. The population tends to build up more rapidly in these brood trees.

Control

In small aspen woodlots or shelterbelts, peripheral heavily infested brood trees may be removed and destroyed before mid-June. Maintaining a heavy understory of other trees and shrubs appears to reduce risk of borer attack. Larvae can sometimes be killed in their tunnels by inserting a piece of flexible wire into the exit holes. No chemicals are currently registered for control of this pest.

Poplar-and-Willow Borer
Cryptorhynchus lapathi (Linnaeus)

Appearance and Life History

The poplar-and-willow borer is a weevil introduced from Europe. Adults are oval, rough, hard-bodied insects about 8 mm long. The body is mottled brown to black with tiny grey or pink scales. A grey or pink posterior portion of the back distinguishes this weevil from most other weevils. Adults have chewing mouthparts at the tips of their long snout. They also feign death (play dead) when disturbed and fall readily to the ground. Eggs are soft, small (1 mm), white and oval. Larvae are white, slightly curved, legless grubs found tunnelling within the bark and sapwood of the host trees. Mature larvae measure up to 13 mm long. Pupae are white to grey or brown, about 10 mm long, with the snout, legs and wings visibly pressed against the body.

The life cycle probably lasts 2-3 years in Canada. Egg laying occurs from July to October on young trees with thin smooth bark. Eggs are laid singly or in groups of two or four in slits chewed by adult weevils in corky bark or scar tissue on stems or limbs. Eggs hatch in 2-3 weeks and the larvae feed on the inner bark, expelling dark brown sawdust out the entrance holes. Larvae may feed from 1 to 2 years, tunnelling in the sapwood, producing girdling and swelling and large quantities of white sawdust that is pushed out of exit holes. When the larvae are full grown, they bore inward and upward and prepare a pupal cell in the center of the stem. Pupation lasts 2-3 weeks. Adults emerge in late summer, mate and begin egg laying shortly after emergence. Adults have wings but flight is rare.

Food Hosts and Damage

Willows of all kinds are the preferred host plants. Some poplars and rarely alder and birch may be attacked if willow is not present. Black cottonwood is the most common poplar host while the trembling aspen is not affected. Borers are commonly found along river valleys where willow grows abundantly. They may become a pest on susceptible trees in landscape plantings.

Minor damage to trees is caused by adult weevils chewing small holes in bark of young shoots. The greatest damage is done by larvae tunnelling one- or two-year-old twigs and stems. Infested trees may be killed or lose their form and become bushy from adventitious growth. Heavily attacked stems are riddled with tunnels and this weakens and causes breakage of stems and limbs. Infestations may be detected by the presence of holes in the bark with sawdust on the bark and on the ground. Oozing of the sap from entrance holes may also be noticed.

Damage to willow by poplar–and–willow borer larvae

M. Herbut

Control

Pruning out and destroying infested limbs and stems in late spring or early summer before the adults emerge and lay eggs is the most practical control. Residual insecticidal sprays applied to the bark in August may control egg-laying adults.

Rose Curculio
Rhynchites bicolor (Fabricius)

Appearance and Life History

The rose curculio or weevil is an occasional pest on both wild and cultivated roses. Adult weevils are about 8 mm long, bright red in color with a black under surface, black head and long black snout or beak. They overwinter as larvae in the soil 3-10 cm deep around the base of food plants. In the spring, the larvae pupate and complete development. Adults emerge in June, about the time roses are in the bud stage. Adults feed on the developing buds, piercing the sepals and petals to feed on the pollen. Mating occurs in July and egg laying begins. The adult female chews holes in the ovary or "hip" of the rose flower, lays a single egg in the hole, and pushes it into the hip and seals the hole with her beak. Small, white, legless curled larvae or grubs hatch and feed inside the hip on the developing seeds. Mature larvae leave the hips in September, drop to the ground and form an earthen cell in which they hibernate over the winter months. Only one generation is produced each year.

Adult rose curculio

H. Philip

Food Hosts and Damage

The feeding action of adult weevils on the buds of both wild and cultivated roses causes failure of the buds to open, or flowers with petals riddled with small holes. Fewer seeds are produced in infected hips. Their presence does not harm the health of the shrubs, but can seriously reduce the number and quality of the blossoms.

Control

Hand removal of the weevils and infested buds will help reduce damage. Picking off and destroying rose hips will get rid of the larvae and reduce the infestation for the following year. No chemicals are presently registered for rose curculio control.

White Pine Weevil

Pissodes strobi (Peck)

Appearance and Life History

The white pine weevil, also known as the spruce weevil, is a native insect found across the coniferous forests of Canada. Adults are dark brown beetles, about 7 mm long, with two elbowed antennae near the top of a prominent, curved snout. The wing covers have irregularly marked bands of reddish-brown and white scales. Eggs are round, 1 mm in diameter, and pearly white. Larvae are legless, curved, creamy white grubs having a small, light brown head capsule with distinctive eye spots. Full-grown larvae may reach 10 mm in length. Pupae resemble the adult but are creamy white at first and gradually become darker as they mature.

White pine weevil larva in damaged white spruce terminal leader

M. Herbut

White pine weevils overwinter as adults in the leaf litter near the base of host trees. They emerge at about the time the buds on the trees begin to swell in the spring and crawl or fly towards the terminal shoots to feed on the bark and mate. This generally occurs in late April or May when the maximum daily air temperature exceeds 10°C. The females feed on the inner bark tissues of the leader and then deposit their eggs in

small punctures in the bark of the leader, just below the terminal bud cluster. A female may lay about 150 eggs singly or in small clusters of three or four eggs. She may lay her eggs on more than one leader and several females may lay eggs on the same leader. Up to 100 eggs may be laid on each leader. Eggs hatch in 1-2 weeks with the larvae boring under the bark to feed on the inner bark, cambium and outer surface of the wood. As they grow, they feed downward towards the base of the leader, girdling the shoot and killing it. Feeding continues for 5-6 weeks until the larvae are mature in about mid-summer. They tunnel into the wood and make small pupal cavities surrounded by wood-chip cocoons plugged with excelsior-like strips of wood removed in excavation. Adults emerge from the infested stems from early August to late September, feed for a period on nearby branches and then overwinter in the leaf litter under the tree. There is only one generation per year.

Food Hosts and Damage

White, Engelmann, Colorado blue, and Norway spruce are the preferred hosts on the prairies. Jack, red, Scots and mugho pine along with black spruce may also be attacked. White pine weevils prefer to feed on open-growing plantations or nursery stock less than 10 metres high and with a leader that is over 12 mm thick. Spring activity can be identified by the presence of weevils or the glistening droplets of oozing resin from feeding punctures below the terminal buds. These feeding punctures allow fungal pathogens to enter the tree. Larval feeding within the shoot cuts off the water supply to the current year's growth. This causes the growth of both the new shoot and needles to slow and yellow before wilting, drying and browning. As the leader wilts, it droops to form a shepherd's crook.

Damage by the white pine weevil, after the leader dies, includes the reduction of marketable timber and the height of the tree. Damage also results in a crook in the stem or a forked tree. One or more of the lateral branches from the living whorl below the killed portion grow upward to become the new leaders. Weevil damage

in ornamental or shelterbelt plantings makes the trees denser and bushier.

Control

White pine weevil damage may be reduced by several cultural methods. Avoid planting host trees near contaminated areas. Susceptible coniferous species may be planted along with fast-growing hardwoods to create a cover of 50 percent shade to make the trees less attractive to the weevils by reducing the thickness of the leader. Avoid planting solid stands of host trees in pure open stands or close to one another. One or two applications of an insecticide to the leader during the flight period of the adults in the spring will prevent oviposition by the females and reduce larval damage. These treatments are effective against the adults only and should only be necessary at 3- or 4-year intervals until the trees are over 10 metres tall and beyond the susceptible stage.

Infested leaders should be hand pruned as soon as the damage is noted, preferably by mid-summer before the weevils emerge. Infested leaders should be cut off below the lowest point of damage and the pruned shoots burned. The strongest branch of the topmost remaining whorl should be trained upward to develop as the new leader and some of the lateral shoots should be cut off. The shape of the tree may be partly saved and only 2 years of terminal growth is lost with this procedure.

Homoptera
Aphids
various species

Appearance and Life History

Aphids, sometimes called plant lice, are fragile, pear-shaped insects measuring 2 mm or less in length. They are generally pale green in color but some are yellowish, brown, pink or black. Adults may be winged or wingless. Most aphids have a pair of tube-like structures known as cornicles protruding from the back. Reproduction is either sexual (mating required) or parthenogenetic (mating not required).

All aphids attacking outdoor broadleaved plants overwinter as small, hard-shelled eggs attached to stems of plants which will serve as food hosts for the newly hatched nymphs in the spring. All newly hatched nymphs are females commonly called "stem mothers". Within a few days they each give birth to live female nymphs. These "daughters" commence feeding on host plants and within a few days these first-generation females give birth to the second generation of females. This reproductive process can continue for 20-30 generations during the summer months, however, no male aphids are produced.

When the aphids become overcrowded or the host plants begin to wilt due to aphid feeding or to lack of moisture, winged females develop and fly to nearby host plants, generally annuals, on which they feed and reproduce for the remainder of the summer. Thus throughout the summer, colonies are continually breaking up and settling on more plants. Some species of aphids feed only on one particular species of plant, whereas other aphids, known as two-host aphids, require two different plant hosts on which to feed and complete their life cycle.

Adult aphids cannot survive prairie winters except on plants maintained in buildings and greenhouses. Cool fall weather initiates a return to winter hosts and the production of winged male aphids. They mate with certain females which lay the overwintering eggs on suitable host plants. Some species of aphids, for example, the grain aphids, do not overwinter on the prairies but migrate from the southern U.S. each spring on southerly winds.

Food Hosts and Damage

Most plants are attacked by one or more species of aphids which may be found on the bark, stem, leaves, blossoms, fruit and roots. Their piercing and sucking feeding behavior produces various abnormal growths such as large, warty growths on the bark, curling and swelling of leaves, and gall formations on foliage and stems. Aphids attacking flowers cause the flowers to wilt and drop off. Severely infested plants will weaken and

eventually dry up.

Various species of aphids are capable of transmitting fungal and viral diseases which can destroy plants, reduce yields or render plant products unmarketable. Several species of aphids attack cereals and legumes, sometimes causing severe economic yield losses (see Grain Aphids and Pea Aphids). Conifers and deciduous trees not only provide overwintering sites for certain aphids, but are also subject to damage by aphid feeding. On broadleaved trees, aphids are generally found feeding on tender, new terminal growth or on the undersides of leaves near the veins. Others cause deformed leaf and bud development which can retard tree or shrub growth. On conifers, aphids feed primarily on buds and newly emerged needles, sometimes causing gall-like growths on the terminals and deformed needles with premature browning and needle drop.

Control

Lady bird beetles and their larvae, lacewing and syrphid fly larvae, and other predators all feed on aphids. Fungal diseases, parasites, high temperatures, damp weather, and hard rains also reduce aphid populations. Various insecticides, both contact and systemic, are available to control aphids. If the aforementioned predators are present on aphid-infested trees, chemical control is not warranted unless the trees are showing severe signs of damage. Control of aphids is one method of reducing the spread of economically important plant diseases caused by viruses.

The green peach aphid, *Myzus persicae* (Sulzer), is probably the most damaging species of aphid attacking field, garden and greenhouse crops. This pale yellow-green aphid overwinters on greenhouse crops or on peach, plum, apricot and cherry trees. Thus, infestations originate mainly from greenhouses or by migrants from the southern U.S. Winged migrants are green with black markings on the abdomen and thorax. Reproduction is continuous in the greenhouse and no eggs are laid. Important host plants include potatoes, cole crops, spinach,

carrots and a long list of greenhouse plants such as carnation, chrysanthemum, dahlia, violet, begonia and fuchsia. Not only do they cause plants to wilt and lose their leaves and reduce their yields, but green peach aphids also are vectors of the viruses that cause such potato diseases as leaf roll, mild mosaic, spindle tuber and various yellows. Viruses that cause yellow mosaic of beans and western yellows of sugarbeets may also be transmitted. Plants infested with green peach aphids may become shiny looking due to the presence of honeydew on the leaves. Production of disease-free stock and control of aphids with insecticides as soon as their presence is noted are the best controls for disease spread. Repeated and thorough applications are sometimes necessary.

Green peach aphid nymph and winged adult

M. Herbut

The potato aphid, *Macrosiphum euphorbiae* (Thomas), is another common aphid on potatoes which overwinters on roses, raspberries and strawberries. This large, slender aphid with curving cornicles is also a vector of many potato diseases and is of concern to potato growers. It attacks the growing tips of plants and deposits large amounts of honeydew on the leaves. Potato aphids also attack tomato, eggplant and lettuce. Aphid control on potatoes includes early planting to produce larger tubers, planting disease-free seed, using a systemic insecticide at time of planting and foliar applied insecticides during the growing season if needed.

The currant aphid, *Cryptomyzus ribis* (Linnaeus), is a small yellow aphid which feeds on the undersides of currant leaves. It overwinters in the egg stage

Potato aphid adult

M. Herbut

on the new growth of currant plants. Eggs hatch as the leaves begin to open, and the young aphids begin to feed on the undersides of leaves, causing puffed red and yellow blisters on the upper surface of the leaves of red, black and golden currants, hedge nettle and occasionally gooseberry. Loss of plant vigor and premature leafdrop occur in severe cases. Honeydew is also evident on fruit. Apply insecticides prior to blossoming for control if infestations have caused damage the previous year.

Black currant leaf damaged by currant aphids

H. Philip

The woolly elm aphid, *Eriosoma americanum* (Riley), is a pest of elm in the spring and saskatoon and roses in the summer. Feeding at the edges of young elm leaves in the spring causes leaves to swell, curl and roll under the edge. Bluish-white woolly aphids will be found inside the leaf curls. Two generations are produced on the elm before these aphids move to a rose or saskatoon shrub to feed on the roots. During the fall, winged adults develop which fly back to the elm to mate and lay overwintering eggs in the cracks of the bark. Damage to trees is minimal but aesthetically unappealing. Honeydew is deposited on leaves and objects under infested trees. Spraying trees with an

insecticide in the spring before the leaves curl may be beneficial, however, once the leaves have curled, chemical control is ineffective.

Woolly elm aphids feeding on American elm leaves

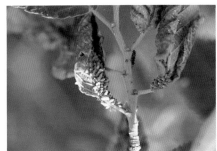

H. Philip

Poplar gall aphids, *Pemphigus* spp., produce closed galls on the leaves, petioles or stems of poplar trees and have an alternate generation on the roots of herbaceous annual hosts such as sugarbeet, lettuce and cabbage, where they can cause serious economic losses. There are several species of poplar gall aphids and each produces a characteristic swelling at a specific site on the tree. They overwinter as eggs near the buds on poplar trees and the nymphs move to newly unfolding leaves where they mature and produce wingless females in the galls. If the plant or leaf is fully grown, galls will not develop. Damage to the tree is not serious and is more a problem of appearance than tree health. Once the galls have been formed, the only control is pruning. An application of insecticide early in the spring may control the aphids and reduce gall formation.

Gall produced by the poplar leaf-stem gall aphid

M. Herbut

The honeysuckle "witches'-broom" aphid, *Hyadaphis tataricae* (Aizenberg), is a recently introduced pest of tatarian honeysuckle, *Lonicera tatarica*, its cultivars, and some related species. The pale green to cream-colored aphids are less than 2 mm in length. They overwinter as eggs or adults in the damaged dried-out leaves on the tips of annual growth. Eggs hatch in spring about the time of active bud development, and feeding occurs on new growth throughout the season. Aphid feeding causes the leaves to fold upwards resulting in deformed gall-like clusters; leaf development ceases and secondary vegetative bud development occurs producing new terminals which rapidly become infested. The ultimate result is a profusion of short, weak terminal stems with very stunted leaves. In late summer, infested foliage dies prematurely. Winged aphids are found throughout the summer which indicates that reinfestation is continuous over the summer. No alternate summer host has been found. The following spring, the infested terminals are found to have been killed, with unsightly "witches'-broom" persisting. Controls include pruning damaged branches in the off-season to remove overwintering eggs, the application of systemic insecticides at 1- to 2-week intervals throughout the summer, and hopefully in the near future, the selection of resistant cultivars.

Honeysuckle damaged by honeysuckle "witches'-broom" aphids

M. Herbut

Cooley Spruce Gall Adelgid
Adelges cooleyi (Gillette)

Appearance and Life History

Several species of adelgids cause needle deformations on spruce. The most common species is the Cooley spruce gall adelgid, often called Cooley spruce gall aphid. Adelgids differ from aphids in structure, physiology and life history patterns. Adelgids produce a terminal gall on primary host trees, undergo as many as six distinct forms, and may require 2 years to complete their life cycles. Adelgid nymphs and adults are both quite small (1 mm in length), light to dark brown in color and covered with white cottony wax.

Cooley spruce gall adelgids overwinter on spruce as nymphs which develop into wingless females or "stem mothers" in the spring. These females are parthenogenetic (reproduce without mating), and lay their eggs near needle buds. Within 2 weeks, the adelgids hatch and commence sucking sap from the base of new needles. Feeding of the stem mothers causes the needle bases to swell and grow around the nymphs, thus forming galls. Nymphs develop into mature wingless females about mid-summer and leave the galls to feed on older needles. Several generations can be produced in one season, however, only the first generation causes galls to form. In late summer, winged females are also produced which fly to an alternate host to establish new colonies. These colonies overwinter on the alternate host and the following summer winged male and female adelgids are produced which mate and return to the primary host to lay eggs. These eggs hatch in 3-4 weeks and the nymphs settle on the stem below the dormant buds to overwinter. Females secrete white cottony wax as a protective cover for their eggs. The appearance of these cottony specks on the tree limbs in early spring and throughout the summer is characteristic of the adelgids' presence.

Food Hosts and Damage

The primary host is spruce with white, Englemann and Colorado blue spruce, being the most commonly attacked. The alternate or secondary host is Douglas fir. The presence of this pest is usuallly recognized by the characteristic galls and white cottony specks and not by the insects themselves.

Cooley spruce gall adelgid damage to white spruce

H. Philip

Although spruce gall adelgids do not harm the health of spruce, they may be a problem in ornamental spruce where their feeding damage reduces the attractiveness of the trees. Adelgids feed by piercing the plant tissue and sucking up the plant fluids. Newly formed galls are 2.5-5 cm in length and gradually change color from light green to dark purple. When the nymphs mature in July or August and leave the galls, the galls turn brown, dry and hard, and can remain on the trees for 2-3 years. It is the presence of these brown galls on the spruce which gives them an unsightly appearance. Adelgids feeding on the older needles can cause the needles to dry out and fall off.

On Douglas fir, the adelgids appear as small cottony specks on the needles. Their feeding causes the needles to bend and turn yellow. Severe infestations will often cause browning and premature shedding of needles.

Control

Removal and destruction of the galls in June before they open can reduce the adelgid population slightly and improve the appearance of the trees. Insecticides may be applied in the spring when the white cottony specks appear to control the stem mothers and again in the fall after the galls open and the white cottony specks appear. It is advisable to avoid planting Douglas fir and spruce near each other. However, galls may be abundant even when Douglas fir is scarce or absent.

Hymenoptera
Birch Leaf-Mining Sawflies
various species

Appearance and Life History

At least three introduced species of birch leaf-mining sawflies attack native and introduced birches in western Canada.

The birch leafminer, *Fenusa pusilla* (Lepeletier), is the most common of the species. Adults are jet black, wasp-like insects, about 3.5 mm long with a 7 mm wing-span. Eggs are ovoid and translucent when laid but gradually turn an opaque white. Larvae are flattened with a brown head and a white body. Four black spots may be found on the underside. Larvae may reach a length of 7 mm. Pupae are white, 3.3 mm in length and gradually turn brown.

Birch leafminers overwinter as full-grown larvae in the soil and pupate in the spring. Adults emerge, mate and begin egg laying in mid-May. Eggs are laid singly in slits on the upper surface of young birch leaves near the midrib. Larvae enter the leaves soon after hatching and feed on the inner leaf tissue. Mature larvae emerge from the interior of the leaf in early July and drop to the ground to pupate in earthen cells 2.5-5 cm below the soil surface. Two weeks later they emerge as adults, and the life cycle is repeated. Usually two or three generations are produced each year on the prairies.

The late birch leaf edgeminer, *Heterarthrus nemoratus* (Fallen), and the ambermarked birch leafminer, *Profenusa thomsoni* (Konow), look similar to the birch leafminer but have only one generation per year and reproduce without mating. The females lay eggs in mature leaves in late June and early

July. The late birch leaf edgeminer pupates in the leaf rather than in the ground.

Food Hosts and Damage

Both native and introduced birch are attacked by birch leaf-mining sawflies. Damage begins as small light green areas on the leaf surface which become noticeably larger until large brown blotches cover parts or all of a leaf. Larval feeding between the upper and lower surfaces of the leaf removes the green tissue to cause these brown blotches. Larvae and frass can easily be found inside the blotches. Birch leafminer lays its eggs on young leaves and near the crown of the tree where new growth occurs. Its damage is early in the season and near the top of the tree. The other species attack older leaves of the trees and their damage is greatest lower on the tree and later in the season.

Birch leafminer larval feeding damage to birch leaf

H. Philip

Egg laying may cause leaf growth to stop or slow down but damage is usually slight. Leaf destruction by the larvae rarely causes infested trees to die. However, the aesthetic damage caused by these pests may be considerable. Entire trees may have their leaves attacked by the later species reducing the attractiveness of the trees. Several years of leaf destruction may reduce growth and vigor of the trees.

Control

Systemic insecticides may be applied to the soil, bark or leaves as soon as damage is observed in the spring. Soil drenches applied in spring near the drip line after the leaves are fully open is the most effective control for season-long control of all species. Soaking the roots in the fall before frost sets in, applying a suitable tree fertilizer each spring, and watering during dry periods in the summer will keep the tree healthy and better able to withstand leafminer attacks.

Dogwood Sawfly
Macremphytus tarsatus (Say)

Appearance and Life History

The dogwood sawfly is an occasional pest on the prairies. The larvae are the most conspicuous and noticeable stage of this species as adults are rarely seen. Larvae have a shiny black head and after the second molt the body is covered with a white powder-like material that can be rubbed off. The early instars are whitish, but when nearly full grown, the back is creamy yellow marked with greyish-black transverse bands or spots and the legs are yellowish. Full-grown larvae are about 35 mm long.

Dogwood sawfly larva

M. Herbut

The color changes of the larvae may aid in their survival. The white coloration of the early instars mimics bird droppings and thus helps to avoid enemies. Likewise the spotted mature larvae are well camouflaged as they crawl over ground litter beneath the bushes.

Adults emerge from late May to July. Eggs are laid on the undersides of leaves with well over 100 eggs deposited on a single leaf. The larvae are present from July to October. When full grown, the larvae wander about, apparently in search of an overwintering site. Winter is passed as prepupal larvae in cocoons in cells constructed in rotting wood found lying on the ground. Pupation occurs in the spring. There is only one generation per year but larvae may remain dormant in cells for one or more years.

Food Hosts and Damage

Dogwood sawfly is a defoliator of dogwood (*Cornus*). Larvae feed in colonies, initially skeletonizing the leaves, then eventually stripping the leaves, damaging the aesthetic appearance of the shrub. Larvae seeking overwintering sites have burrowed into composition wood fiber wallboard and into wooden furniture set on the ground.

Control

Chemical control of dogwood sawfly is only necessary if ornamental plantings are being noticeably defoliated. Picking off and destroying the larvae should be sufficient if they are not numerous.

Pear Slug
Caliroa cerasi (Linnaeus)

Appearance and Life History

The pear slug, or pear sawfly, is an introduced insect from Europe. The black and yellow wasp-like adult sawflies are about 10 mm long, shiny, rather stout-bodied with clear wings. The larvae resemble tiny slugs and are coated with a dark olive-green to yellowish slimy covering. The front end of the larva is wider than the rest of the body which may reach 12 mm in length. The pupae form in earthen cells in the soil.

Pear slug larva feeding on cotoneaster leaf

H. Philip

Pear slugs pass the winter in cocoons 5-7 mm below the soil surface or in leaf litter beneath the host plants. In late spring, shortly after the host plants come into full leaf, adult sawflies emerge from these cocoons and fly to the host plants where mating takes place. Females insert their eggs singly into the leaves and a tiny blister forms over the place where each egg was laid. Eggs hatch in 9-15 days into soft-bodied "worms" or "slugs" which begin to feed on the upper surface of the leaves. The feeding period varies from 2 to 3 weeks and as the slugs grow in size they become somewhat lighter in color. When full grown, the yellow larvae drop or crawl to the ground to pupate inside cocoons. Adults emerge again in late July and early August. It is the second generation that usually causes the greatest amount of injury during the late summer and early fall. When the larvae of this second generation are full grown, they drop to the ground, pupate, and remain there for the winter. There are two generations per year on the prairies.

Food Hosts and Damage

Cotoneaster, hawthorn, mountain ash, cherry, plum and pear are the trees and shrubs most commonly attacked. Pear slugs do not chew the whole leaf, but skeletonize the upper surface tissue leaving the leaves with scorched, brownish areas interlaced with leaf veins. Severely damaged leaves fall off prematurely. This can seriously affect fruit yield and the aesthetic beauty of ornamental plants. High populations can lead to discoloration and defoliation of entire cotoneaster hedges and individual fruit and mountain ash trees. More damage by pear slugs is noticed in late summer as the second brood of larvae is generally more numerous than the first or early summer brood. Healthy trees and shrubs should be able to withstand several years of moderate attack because the damage occurs in late summer when the growing season is nearly complete.

Control

Susceptible trees and shrubs should be checked frequently in mid-June for first-generation larvae and from mid-July onward for second-generation pear slug larvae or evidence of feeding. Infested leaves of single trees or hedges may be blasted with a strong stream of water

from a garden hose, or with a soapy solution from a watering can. A contact insecticide applied when the larvae are first noticed will also control this pest.

Webspinning Sawflies
Cephalcia spp., *Acantholyda* spp.

Appearance and Life History

Webspinning sawflies are intermittent pests of spruce and pine in western Canada. The wasp-like adults emerge from the soil in the spring and the females lay small green eggs singly or in rows of two to four on the needles. After hatching, the larvae feed in groups, spinning silken nests which are filled with discarded food, cast skins and frass. The nest is formed in the crotch of a twig and a branch. The larvae resemble other sawfly larvae with three pairs of thoracic legs but they lack abdominal prolegs. Larvae also have many-segmented antennae, a pair of jointed structures on the last abdominal segment, and a flattened abdomen. Full-grown larvae may be about 25 mm long with a black head and thorax and greenish-brown body. When mature, the larvae drop to the ground and overwinter in earthen cells below the host trees. Pupation occurs in the spring and adults emerge soon thereafter.

A webspinning sawfly larva

H. Philip

Food Hosts and Damage

Webspinning sawflies feed on various spruces and pines across the prairies. Larvae feed by cutting old needles and taking them back to their nests constructed of needles and frass bound together with webbing. Damage is of most concern to ornamental plantings as the cut needles and nests at the base of the twigs give an unsightly appearance

to the tree. Nests may be 5-10 cm in diameter with the needles on adjoining twigs completely stripped off.

Control

Control may be required on ornamental plantings. Nests may be removed by hand and destroyed if they still contain larvae or they may be knocked out of a tree with a hard stream of water. A contact insecticide applied with sufficient force to penetrate the webbing and frass may be beneficial.

Willow Redgall Sawfly
Pontania proxima (Lepeletier)

Appearance and Life History

Adults of the willow redgall sawfly are small (3.5-5 mm long) shiny black wasp-like insects. Larvae are also small (5 mm long), pale-colored with a dark head, with true legs present on the thorax and prolegs on each abdominal segment.

Adults emerge in late spring, mate, and seek out suitable host plants in which to deposit eggs. Females insert their eggs in the leaf tissue and within a few days the eggs hatch and the small larvae commence feeding on the inner leaf tissue. Larvae leave the galls to drop to the ground and pupate in mid-summer. The second brood of adult sawflies repeats the cycle and the subsequent larvae pupate in the fall and overwinter in this stage. There are normally two generations produced annually.

Food Hosts and Damage

Willow is the only host. The feeding activity of the larvae causes the formation of leaf galls which are bean-shaped, smooth, and stick up equally on both sides of the leaf. Galls may be green, yellow or red. A single larva feeds in the cavity of each gall. The galls produced on the leaves do not harm the trees, but do reduce their attractiveness.

Willow redgall sawfly galls on a willow leaf

H. Philip

Control

Galls can be pruned off as they form and leaves may be collected and burned in the fall to reduce overwintering populations. Some control of the early generation has been reported with insecticidal sprays on newly developing leaves. However, chemical control is not usually necessary or recommended.

Willow Sawfly

Nematus ventralis Say

Appearance and Life History

The willow sawfly has been occasionally abundant on the prairies. Adults are medium-sized sawflies that are rarely seen. Larvae are 20-24 mm long, near black with 11 conspicuous yellowish spots on the sides of the body. They have six pairs of greenish-yellow prolegs.

Willow sawfly larva feeding on aspen leaf

Department of Entomology, University of Alberta, Edmonton

Winter is spent in the prepupal stage in cocoons in the litter or top soil beneath host trees. Pupation and adult emergence occurs in the spring. Eggs are deposited in pockets cut in the leaves. Young larvae feed in colonies. A larva rests or may feed in a "J" position with the tail end bent upward. One to several generations may occur during the summer. Mature larvae drop to the gound to overwinter. Some individuals may remain dormant up to 20 months before completing their development.

Food Hosts and Damage

Willow sawflies can completely defoliate willows in ornamental plantings and along streams. They also attack new growth of poplar. Initial damage appears as small holes in the leaves; as the larvae mature, more leaf tissue is consumed until partial or complete defoliation results, depending on the number of larvae present.

Control

Predators and parasites generally are effective in preventing or reducing willow sawfly damage, thus chemical control is rarely warranted. Damage to trees is usually not serious and has no long-term effect on the growth of the trees. Chemical sprays should be applied to trees that are heavily infested as soon as the damage is considered severe.

Yellowheaded Spruce Sawfly

Pikonema alaskensis (Rohwer)

Appearance and Life History

The yellowheaded spruce sawfly is a native North American insect pest of spruce. Adults are yellow to reddish-brown wasp-like insects about 10 mm long, with four transparent wings and a saw-like ovipositor for cutting slits in the needles within which the eggs are laid. Eggs are minute, about 1 mm long, blunt, spindle-shaped and pearly white. Newly hatched larvae have a yellowish head and a light yellowish-green hairless body. Mature larvae are about 20 mm long, the head distinctly yellowish-brown, the body olive green on top and pale beneath with a series of dark stripes along the back and sides. Larvae have three pairs of legs on the front portion of the body and seven pairs of fleshy feet or prolegs on the abdomen. When disturbed, the larvae exude a liquid from the mouth and they arch both ends. Pupae are enclosed in an oval, light brown to reddish-brown papery cocoon about 12 mm long encrusted with soil particles and other debris.

Yellowheaded spruce sawfly larva feeding on white spruce

H. Philip

Adult sawflies emerge from overwintering cocoons in late May and early June about the time new terminal growth appears on spruces. Females deposit their eggs in the new needle growth and larvae hatch in 1-2 weeks. They mature in 3-8 weeks, feeding in groups on the new needle growth until only short, brown stubs remain. Mature larvae drop from the needles to the ground where they burrow in the lower duff layer or mineral soil to spin cocoons. They overwinter as larvae and pupate in the spring. Only one generation is produced annually.

Food Hosts and Damage

Yellowheaded spruce sawflies can be found wherever spruce grows in North America. They prefer young, open-grown trees, especially those grown in shelterbelts, plantations, hedges, roadside plantings and ornamental trees growing singly. Damage to spruce is generally not noticed until the larvae are quite large and devouring the needles. They begin by eating only parts of new needles, then eat all the new needles before moving onto old ones. This stripping of needles and yellowing of the tree starts near the top of the tree and gradually moves downward. The loss of needles can retard tree growth and repeated attacks by this pest can seriously weaken the tree making it more susceptible to further damage by other pests, to competition from other trees and to adverse weather conditions. Two or 3 consecutive years of moderate defoliation is enough to kill a tree.

Control

Single trees or small groups of trees may be protected by hand picking the larvae when they first appear and destroying them. Another effective method is to hose the larvae off with a garden hose. On large trees or on many trees, chemicals applied in mid- or late June to the larvae should give adequate protection to the trees. Natural predators and parasites generally are unable to protect susceptible trees under weather conditions favoring sawfly development.

Lepidoptera
Ash Borer
Podosesia syringae (Harris)

Appearance and Life History

The ash borer is an occasional pest across the southern prairies. Adults are day-flying, wasp-like moths with transparent hind wings, slender abdomen with yellow bands, and very long, yellow and black hind legs. They have dark brown or black bodies, and a wing-span of 25-35 mm. Eggs are tan, elliptical in shape and 0.7 mm long. Newly hatched larvae are white with an amber-colored head. Mature larvae are about 26 mm long, creamy white with a shiny brown head. Pupae are tan to reddish-brown, 18-24 mm in length, and have small backward-projecting spines on the abdomen.

Moths emerge in early summer from May through July and after mating the females lay their eggs on bark, in cracks and crevices or in tree wounds. Ten to 14 days later the larvae hatch and bore into the bark. Entrances to new galleries are marked by a mixture of fine boring dust, oozing sap and excreta. During the first summer, larvae feed within the bark and may bore into the wood later in the season. Some larvae overwinter in the bark, others in the sapwood. During the second summer, they bore into the wood as they feed. Larvae spend the second winter in the heartwood. In the spring the larvae tunnel close to the bark surface and pupate. The pupae break the thin layer of bark and push themselves partly out the exit hole just prior to adult emergence. When the moth emerges, the pupal skin is left in the exit hole. One generation takes 2 years to complete.

Food Hosts and Damage

Green ash and lilac are the preferred hosts of the ash borer, with Mountain ash and privet as alternate hosts. Scars, accompanied by enlarged or swollen areas, are associated with repeated infestations. Branches may be weakened at the feeding sites. Holes and burrows, about 6 mm wide, may be found in trunks and branches with boring sawdust surrounding and beneath the exit holes. Heavily infested trees and branches may wilt during hot days.

Ash borer damage to green ash

Northern Forestry Centre,
Canadian Forestry Service, Edmonton

Control

This pest is attacked by predators and parasites. Tree care to improve tree vigor may reduce borer survival and enable trees to withstand attack. Cover fresh wounds with paint or wound dressing to prevent further bleeding and entrance of disease-causing organisms. Remove and burn old, dying and infested trees or shrubs damaged beyond recovery. If only one or two trees are infested, wrap the trunks and basal parts of the main branches in mid-May with burlap or cotton cloth to trap the emerging moths and remove the wrapping in mid-August. Repeat annually for 3 years.

Bedstraw Hawkmoth
Hyles gallii (Rottenburg)

Appearance and Life History

The bedstraw hawkmoth is a large moth with long narrow front wings and small round hind wings. It has a wing-span of 4-7 cm and a body length of 3.5 cm. The front wings are a shade of greenish-brown with an irregular but prominent greyish band from tip to base. The hind wings are dark brown with a broad pink band across each wing. The stout, hairy body tapers to the end and has six white lines on the thorax. The antennae are thickened and somewhat spindle shaped. The larvae are large (up to 6 cm long), hairless caterpillars with a red head and a red horn on the hind end of the body. Larvae vary in color from red brown, brown to black, or greenish-grey, with light colored spots along the sides. Legs and prolegs are black.

Bedstraw hawkmoth larva

M. Herbut

Moths fly from late May into the summer and are common around lights at night. Larvae feed during the summer and pupate in the soil in late summer when they are full grown. They overwinter as pupae and only one generation occurs each year.

Food Hosts and Damage

Larval food plants include bedstraw (Galium spp.), fireweed (Epilobium spp.) and annual Clarkia. When they have consumed these weeds, they may move out of the roadside ditch and shelterbelt in search of other host plants.

Control

There are no registered chemical controls for bedstraw hawkmoth larvae since they rarely do sufficient damage to warrant control.

Boxelder Twig Borer

Proteoteras willingana (Kearfott)

Appearance and Life History

.
The boxelder twig borer is the larva of a small grey moth with streaks and clusters of yellow to black scales on its wings. The wing-span varies from 15 to 20 mm. Pearly white eggs are round and flattened with a flange-like edge. Newly hatched larvae are yellowish-white, about 1 mm in length with a light brown head capsule. As larvae mature, they gradually change to a more yellowish-green body color with dark brown to black head capsules. Larvae reach a length of 12 mm when mature. They pass through several larval stages before pupating. Pupae are reddish-brown, 7-10 mm long.

Boxelder twig borer larva feeding on boxelder

Northern Forestry Centre,
Canadian Forestry Service, Edmonton

Moths fly in late June and July during evenings and on warm cloudy days with a rapid, darting motion. Moths may rest on three trunks and on the ground but not on the leaves. Females lay eggs singly on the underside of leaves near the midrib or larger veins. They may lay up to 100 eggs each during their flight period. Eggs hatch in about 2 weeks during mid-July to early August and newly hatched larvae feed on the leaves until September. Larvae then bore through the base of a leaf stem and into dormant buds where they overwinter. In the spring, they emerge from their overwintering buds and resume feeding. In late May, the larvae leave the buds and bore into the new tip growth on the branches with burrows reaching 3 cm or more in length. By early June, the mature larvae leave the galls, drop to the ground and pupate in cocoons among the leaf litter on the soil surface. Adult moths emerge in 2-3 weeks. There is only one generation per year.

Food Hosts and Damage

Manitoba maple on farms and urban areas is the only known host on the prairies. Parts of the leaves near the midrib and veins are eaten in July and August. Dormant leaf buds are bored into and destroyed. Each larvae may destroy two or three buds. Infested terminal growths resemble swollen and spindle-shaped galls. Infested twigs dry out, become woody and die. This prevents further terminal growth, promoting secondary branching and bushy trees. Damage is generally considered minor.

Control

In most years, boxelder twig borer damage is not extensive enough to warrant control. Natural predators and parasites keep it under control. Cutting off the twig galls in late May and early June when the larvae are still in them will reduce future populations. Removal of the suckers around older trees helps remove hibernating larvae and burning them destroys the borer population. Contact insecticides may be useful if applied in mid-July to early August to control the young larvae before they move into the buds or twigs.

Bruce Spanworm

Operophtera bruceata (Hulst)

Appearance and Life History

The Bruce spanworm, a native insect, is a periodic pest of aspen and willow on the prairies. Adult males are slender medium-sized moths, 7-10 mm long with a wing-span of 25-30 mm. The semi-transparent forewings have light brown and grey bands on a brown background. The hind wings are uniformly beige. Adult females are 6-7 mm in length but lack wings. The thick body is dull brown covered with irregular patches of white scales giving it a rough appearance. Eggs are oval in shape and covered with deep pits. Newly laid eggs are pale green when laid, change to bright orange over the winter and turn grey in the spring. Larvae are smooth hairless caterpillars with three pairs of true legs on the thorax and two pairs of prolegs on the abdomen. Head and body color vary from bright green to dark brown with three white lines on the sides. Mature larvae may be up to 18 mm in length. The 6-7 mm long pupa are golden-brown to green, becoming a dark grey-brown and enclosed in a cocoon covered with grains of sand or humus.

Bruce spanworm larva feeding on aspen leaf

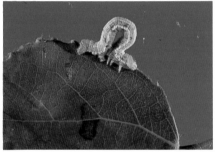

Northern Forestry Centre,
Canadian Forestry Service, Edmonton

Bruce spanworms overwinter as eggs laid singly in bark crevices, under loose bark or in any other protected location on the lower part of the tree. Hatching coincides with bud burst in the spring and larvae begin to feed immediately. If food is scarce, larvae will drop down on a silk thread to seek food hosts or be carried to other food sources by the wind. Larval feeding may last 5-7 weeks in May and June. When mature, larvae drop to the ground and spin individual cocoons 5-10 cm below the soil surface where they remain until fall when they emerge as adults. Adults are active in October and November, after the leaves have fallen and when snow may cover the ground. Males are active in late afternoon or early evening and may be seen flying around the trunks of leafless trees in search of females. Females climb the trees, mate and lay eggs on the trunk of the lower part of the tree.

Food Hosts and Damage

Bruce spanworms feed mainly on trembling aspen and willow on the prairies but a very wide range of broadleafed trees may be attacked. Larvae feed on the buds and leaves of the trees in spring, eating the green matter without touching the ribs. Feeding is usually done on the undersurface of the leaf and may go undetected. During outbreak years, total defoliation may occur in June with re-leafing occurring later in the summer.

Control

Several parasites and diseases generally keep this insect species under control and outbreaks are usually short-lived. The use of a sticky band about 12-15 cm wide around the base of the tree will prevent adult females from climbing the tree to lay eggs. The use of a contact insecticide while the females are climbing is also effective. A well-timed spray when the buds are breaking and the eggs are hatching will prevent serious defoliation by the larvae.

Carpenterworm

Prionoxystus robiniae (Peck)

Appearance and Life History

The carpenterworm attacks many hardwood trees on the prairies. Female carpenterworm moths are greyish and stout-bodied with a wing-span of 75 mm. The forewings are mottled grey and black, with the hind wings a smoky-grey. Male moths are smaller with a wing-span of only 50 mm and a less stout, shorter body. The hind wings are yellowish-orange with a black base and margins. Eggs are oval, olive-brown in color and 2.4 mm long. The shell is hard with a net-like pattern of ridges. Newly hatched caterpillars are dark brown, slightly hairy with large black heads. Full-grown caterpillars are 50-75 mm long, pinkish-white with brown heads and several short, stout hairs on each body segment. Pupae are brown, 30-50 mm long and have circular bands of backward-pointing dark spines.

Carpenterworm larva feeding inside host

Northern Forestry Centre,
Canadian Forestry Service, Edmonton

Carpenterworms require 3 or more years to complete their life cycle. Moths emerge in early June and may be present until early August. Females are poor fliers but males are strong fliers and move about readily. Mating and egg laying takes place soon after emergence with a single female laying 300-400 eggs in small groups of two to six in cracks, crevices and wounds in the bark, or in or near old burrow openings. Eggs hatch in 10-16 days. Larvae crawl about freely on the host trees before they burrow through the bark and into the host. Small shelters, made of particles of frass held together by bits of webbing, are constructed by the young caterpillars to cover themselves and their burrow entrances. By August, frass ejected from the new burrow becomes noticeable. The larval period extends from July of the year in which the eggs are laid to May of the fourth or fifth year.

During the second and following seasons the burrows are extended and enlarged, usually forming a maze of criss-cross tunnels in the wood of heavily infested trees. During this time the caterpillars eject a great deal of coarse frass from the burrows which may cling ribbon-like on the outside of the trunks or accumulate like sawdust around the base of the trees. The caterpillars also return at intervals to the outer sapwood to feed. Often they emerge from the burrows to crawl about on the bark and then re-enter the burrows. In May of the final year, the caterpillars retreat to the upper parts of the burrows where, in specially prepared chambers, they transform into pupae. The pupal period is fairly short as adults begin to appear in June. Before changing to the adult form,

the pupae move downward by means of the backward-projecting spines on their bodies to the burrow openings so that the moths, upon emerging from their pupal skins, escape directly to the outside.

Food Hosts and Damage

Carpenterworms attack green ash, mountain ash, poplar and elm most commonly on the eastern prairies. Willow, maple, cottonwood and fruit trees may also be attacked. They may be serious pests of trees planted for shade, ornamental and windbreak purposes as they prefer sunny exposures. Open-growing plantings such as shade and ornamental trees in parks and on boulevards are more seriously affected. Carpenterworms devour portions of the inner bark and the outer sapwood, thus producing dead, sunken areas on the trunks and branches, frequently resulting in dead tops. The aesthetic value of affected shade and ornamental trees is lost or greatly reduced. Tunnels in the wood weaken the trees structurally and increase the hazard of breakage from winds. Tunnels also permit moisture and decay-causing organisms to enter and cause further deterioration.

Trees infested with carpenterworm may be recognized by small holes 3-25 mm in diameter extending into trunks or branches resulting from larval burrowing. The holes may be few and scattered or they may occur in groups, frequently at the base of the tree or at the lower branch levels. Masses of frass clinging to the trunks or branches, or occurring on the ground at the base of the trees are also indicative of attack.

Control

Physical controls to save and protect individual trees include wrapping the trunk where borer holes occur with burlap cloth in mid-May to prevent the moths emerging from the burrows and laying eggs. This must be continued for 3 or 4 years to ensure all adults developing from the caterpillars occurring in the tree have been destroyed. Caterpillars may be dug out with a pointed knife or may be killed by probing the burrows with a wire. Burrows may be injected with an insecticide to destroy the caterpillars. Trunks and main branches of the host trees may be sprayed with a suitable insecticide to kill the newly hatched caterpillars before they form burrows, and the older caterpillars which leave the burrows during their developmental period to crawl about temporarily on the outside of the trees.

Carpenterworm damage may be reduced by avoiding wounding trees and by covering wounds with commercial tree dressing to reduce desirable egg-laying sites. Brood trees should be removed and destroyed before June to reduce the number of potential egg-laying female moths.

Eastern Spruce Budworm
Choristoneura fumiferana (Clemens)

Appearance and Life History

The eastern spruce budworm is a minor native pest of softwoods on the prairies. Adult budworms are small grey-brown moths with speckled white forewings. Adults are about 13 mm long with a wing-span of 22 mm. Eggs are light green, about 1 mm long by 0.2 mm wide. Eggs are laid overlapping one another to form a flattened silky mass containing about 20 eggs. Newly hatched larvae are about 2 mm long, yellowish-green in color with a light- to medium-brown head. As the larvae mature, body color changes from pale yellow to dark brown with light colored spots along the back. Full grown larvae are 2.5 cm long with a dark brown or shiny black head. Initially pupae are pale green but gradually change to reddish-brown with darkened bands and spots.

Eastern spruce budworm larva feeding on spruce

M. Herbut

Eastern spruce budworms have one generation per year. Egg masses are laid in late June and early July on the underside of needles on shoots exposed to sunlight. Eggs hatch in about 10 days, and larvae disperse by crawling within the crown of the tree or from tree to tree on silken threads carried by wind. These first-instar larvae then spin silken hibernation shelters in crevices of twigs and bark where they molt to second-instar larvae and remain until the following spring. Overwintered larvae emerge in early May when maximum daily temperatures reach 17°C and mine old needles or may again be dispersed by the wind to other trees. As the new buds begin to burst, the larvae begin feeding on the needles within the bud and form protective shelters of silk webbing, bud scales, and debris. Larvae molt five times as they grow, and after each molt they establish new feeding sites and shelters by webbing the needles together. Pupation takes place in mid- to late June. A larva attaches itself either to dead needles and waste material by a silken thread, or to another support by its abdominal tip and transforms into a pupa. Moths emerge in about 10 days, mate and disperse to repeat the life cycle.

Food Hosts and Damage

Eastern spruce budworm prefers white spruce and balsam fir, however, black, Engelmann and Colorado blue spruce, pine and larch may also be attacked. It prefers to feed on the mature and overmature mixed stands of spruce and balsam fir but can survive on pure spruce stands. Even though outbreaks of this insect in eastern Canada may be very severe and long lasting, outbreaks on the prairies are fairly local and short-lived.

The first signs of damage occur in early June when the last-instar larvae feed on the new foliage at the top of the tree. Needles and tips turn a reddish-brown as the larvae feed first on the buds and developing shoots and then on the needles. In severe infestations, all the new foliage, together with some of the older needles, may be eaten for several consecutive years, causing trees to lose their vigor. In 3 years the tree top dies,

and in 5 years the tree dies, depending on the size, vigor and health of the trees attacked. Reduction in tree height and diameter have also been recorded along with reduced cone production from direct cone feeding and loss of vigor.

Control

Larvae can be removed by hand from small trees early in the season and destroyed. Larger trees will require spraying with an insecticide about mid-May when the larvae first appear and before damage becomes noticeable from a distance.

Fall Cankerworm
Alsophila pometaria (Harris)

Appearance and Life History

The fall cankerworm, a native North American insect, may occasionally be a serious pest of hardwood trees on the prairies. It gets its name from the moths which fly only in the fall. The male moths have brownish-grey forewings with two irregular light bands; hind wings are greyish to light brown with tiny spots on the outer part. They have a wing-span of about 30 mm. Female moths are wingless and dark grey in color. Eggs are brownish-grey and shaped like tiny flower pots. They are laid in compact masses on the bark of tree trunks and branches. Larvae are elongate, up to 25 mm long, and vary from light green to brownish-green with a darker stripe down the back. Green larvae may have faint white lines running down the back. The larvae are also known as loopers or inchworms. The pupae are oval in shape, brown, and enclosed in tough cocoons with particles of soil interwoven with silk.

Fall cankerworm larva

M. Herbut

Moths emerge from cocoons in the soil in late September and October, usually after the first heavy autumn frosts. Males appear first and are active mainly at dusk when they may be seen flying around trees in search of females. The wingless females must crawl up the host plant. Mating occurs in the trees and the females lay their eggs in compact masses of about 100 eggs on the twigs, branches, and trunks. The eggs overwinter on the tree and hatch the following spring near the end of May. Larvae usually go through five or six instars during their 5- to 6-week feeding period. During this feeding period, the larvae may spin a silken web and drop down on the strands to avoid predators, to disperse from tree to tree or to lower themselves to the ground. When larvae are full grown, they drop to the ground and burrow through the litter layer and spend the pupal period in a cocoon 3-10 cm in the soil. There is one generation of fall cankerworm per year on the prairies.

Food Hosts and Damage

Fall cankerworm larvae feed on most broadleafed trees and shrubs, but prefer American elm and Manitoba maple in shelterbelts and urban settings. Basswood, Siberian elm, apple, ash and poplar along with lilac, honeysuckle and roses are also attacked. The first noticeable sign of an infestation is the appearance of small "shot-holes" in the young leaves. At this time the tiny larvae are found on the underside of the leaf. As the larvae continue feeding the holes grow larger until almost all of the leaf tissues are eaten. During severe outbreaks trees and shrubs may be completely defoliated. Healthy trees and shrubs usually produce a new crop of leaves by mid-July and show little permanent injury from a single defoliation. However, after 3 or more consecutive years of heavy attack, tree growth is slowed down and branches in the crown die back. An epidemic may last 3 or 4 years and then may not be a problem for another 15-20 years. Many larvae dropping to the soil during years of high populations may be a nuisance in urban or recreation areas.

Control

Natural controls by parasites, predators, starvation and weather play the biggest role in controlling fall cankerworm populations. Painting a band of some sticky substance around the tree trunk in mid-September will prevent adult females from climbing up the host plant to lay eggs. Bands must be 7-10 cm wide and 1-2 metres above the ground. Contact or stomach insecticides may be used against the larvae in the spring. The use of bacterial insecticides are beneficial if applied when the larvae are quite small.

Forest Tent Caterpillar
Malacosoma disstria Hübner

Appearance and Life History

The forest tent caterpillar is a native insect found wherever hardwoods grow throughout North America. Adult forest tent caterpillars have stout hairy bodies, rusty-red to a pale fawn in color, with a wing-span of 20-40 mm. The forewings have two narrow, pale, oblique transverse lines bordering a wide dark band. Males are much darker and usually smaller than females. Eggs are grey, shaped like a flower pot and are laid side by side in a band around small twigs. The egg bands are covered by a brown foam-like protective covering. Bands may be 12-35 mm wide and contain 150-350 eggs. Young larvae are

Forest tent caterpillar egg band

M. Herbut

uniformly black with distinct dark hair but colors gradually appear as the larvae mature. Mature larvae, measuring 45-50 mm in length, are dusky brown with four yellowish-brown stripes extending the length of the body, a row of diamond or

key-hole shaped white spots down the centre of the back, a broad band of blue along each side, and covered with long, fine brown hairs. Pupae are black to brown and enclosed in yellowish silken cocoons constructed in folded leaves, bark crevices, under eaves, or other sheltered places.

Forest tent caterpillar larva

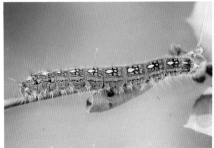

M. Herbut

Forest tent caterpillars overwinter as fully developed caterpillars within the eggs and emerge when leaves begin to unfold in the spring. Larvae feed on the leaves of host trees for 5 or 6 weeks. They do not congregate in tents but cluster on silken pads and spin silken paths along which they travel on the trees. They cluster on tree trunks or branches when resting or when weather conditions are unfavorable. Just prior to pupation in late June or early July, the caterpillars begin seeking suitable sites to spin their cocoons and pupate. After 10 days the adults emerge and begin flying. During their 5- to 10-day life span, adults mate and lay eggs for the following year's generation. There is only one generation per year.

Food Hosts and Damage

Forest tent caterpillars are major pests of aspen poplar as well as birch, Saskatoon, green ash, apple, mayday and most other species of broadleaved trees and shrubs. The rate of defoliation increases as the caterpillars grow larger. Thousands of hectares of aspen forest can be defoliated during an outbreak in any one area. The outbreaks, lasting 3-6 years, occur on the average every 10 years but non-outbreak intervals may be as short as 6 and as long as 16 years. Severely defoliated trees rarely die unless attacked 3 or more years in succession. Usually there is reduced annual growth and some branch and top mortality.

Control

Outbreaks of forest tent caterpillars are usually terminated by a combination of starvation, parasites, predators, diseases and unfavorable weather conditions. On ornamental trees, removing egg bands on the branch tips by pruning or scraping is effective. This can easily be carried out in the fall or spring when the shrubs are bare and egg bands are more visible. Once the eggs have hatched, the larvae can be picked off and destroyed when they cluster together in the evening or on cool days. Chemicals may be applied in the spring to control larvae if damage is severe.

Glassy Cutworm
Crymodes devastator (Brace)

Appearance and Life History

The glassy cutworm is a sporadic pest across the prairies. Adult moths have dark grey forewings with a mixture of light grey and blackish markings, and two light grey spots near the center of each wing. The hind wings are light grey becoming paler toward the base. Wing-span is 30-40 mm. Larvae have a translucent, greenish-white smooth body with a reddish-brown head and neck shield. Mature larvae measure 35-40 mm in length. Pupae are brown, spindle-shaped and about 30 mm in length.

Glassy cutworm larva

M. Herbut

Moths fly from June to early September. Eggs are laid near the base of grasses and hatch soon after. Glassy cutworms overwinter as partly grown larvae, feeding in the fall and spring, and pupate

in the soil before the end of May. The larvae feed exclusively underground and never come to the soil surface. There is one generation per year.

Food Hosts and Damage

Glassy cutworms are most commonly associated with grasses and turf but may attack cereals, corn and vegetables planted on newly broken sod or in crops with heavy grassy weed populations. Timothy, bluegrass and brome pastures are commonly infested but damage is rarely noticeable. Damage to turf and lawns can be more serious and noticeable as the cutworms clip off the leaves and pull the plants underground to feed, causing patches of bare soil or dead grass in the turf.

Control

Since glassy cutworms prefer to lay their eggs in or near the base of grasses, land that is to be broken should be well cultivated to cover the sod. Similar care should be taken on grassy summerfallow. Controls in turf are similar to those for sod webworm.

Large Aspen Tortrix
Choristoneura conflictana (Walker)

Appearance and Life History

The large aspen tortrix may periodically be a serious pest of trembling aspen. Adult moths are brownish-grey with indistinct dark grey markings on the forewings and solid grey hind wings. The wing-span may be up to 35 mm, making it the largest tortrix in North America. Eggs are oval, scale-like and pale green, and are laid in a flat cluster or patch on the upper surfaces of leaves. A cluster may have 50-450 eggs. Young larvae are pale yellow with light brown legs and head capsule. As the larvae grow, they become darker green to nearly black in color with a dark brown to black head and a reddish-brown to black shield on the back just behind the head. Full-grown larvae measure about 20 mm long. The spindle-shaped pupae change from bright green to reddish-brown over the 2-week pupal stage. Pupae are 13-17 mm in length.

Large aspen tortrix larva

M. Herbut

The large aspen tortrix has only one generation per year. Adult moths fly, mate and lay eggs from mid-June to mid-July. Female moths are slow moving and do not generally fly much while the males are very active flyers. Eggs hatch in about 2 weeks and the young larvae spin webs, tying two or three leaves together and rolling them up. Larvae are gregarious and prefer to feed as a group, skeletonizing the leaf surfaces. Many larvae hang by a silken thread and are transported by the wind, thus extending their range. In August the larvae move to protected sites in cracks and crevices of the bark, often at the base of trees, in moss, or beneath dead bark on twigs on the ground, and spin silken cells known as a hibernacula. The larvae molt to the second instar and remain dormant for the rest of the winter.

In the spring, just before the buds begin to break, the larvae crawl up the tree on warm days, often building a second shelter. Once the buds break, the larvae penetrate inside the buds to continue their development. As the leaves open, the larvae tie them together and roll them up to feed inside. Feeding continues until the larvae are mature in early June. During outbreak years when all the leaves have been stripped, the larvae may search for more food, dropping on webs from the tree causing extensive webbing on grass and underbrush. Pupation usually occurs within the leaf shelter on the tree or on the undergrowth if the trees have been stripped of their leaves.

Food Hosts and Damage

The large aspen tortrix is found wherever trembling aspen grows. During severe outbreaks, it may also attack balsam poplar, white birch, willow and alder. The damage caused by first-instar larvae the first summer is minimal and generally goes unnoticed. Damage in the spring on the new buds and leaves can be quite serious. New buds may be completely destroyed, preventing leaf development. Rolled leaves and loss of leaves detracts not only from the appearance of ornamental trees but reduces the annual growth of the trees. Trees are rarely killed. Healthy trees usually grow new leaves by mid-summer, often smaller and fewer than normal, causing thinned crowns. Repeated defoliation over several years may cause trees to be weakened, making them more susceptible to other factors such as drought and disease. Outbreaks usually only last 2 or 3 years.

A large aspen tortrix infestation can be detected by the delayed opening of the buds in the spring. The deformed leaves and clusters of leaves webbed together will also aid in identification of this pest.

Control

A number of natural enemies help to keep large aspen tortrix populations under control. These include parasitic flies and wasps and predators such as ants and birds. Viral and fungal diseases have helped reduce larval populations. Cool, wet weather may not only increase disease incidence, but will also slow up development of the larvae while the trees have the best growing conditions. Frost in the spring when the larvae are coming out of their overwintering sites may kill some larvae.

The use of a sticky band around the base of trees has been recommended to stop the marching stage in late summer and again early in the spring. The use of contact insecticides may be necessary to protect high-value trees. Penetration of the webbing and spraying the tops of trees may be difficult. In most instances, no controls are necessary.

Leafrollers and Leaftiers
various species

Appearance and Life History

There are many species of small caterpillars that roll, fold or tie leaves of trees and shrubs together. Their appearance and life cycles vary considerably. Adults of the most common family of leafrollers (Tortricidae) are small moths with wing-spans under 25 mm with the forewings mottled yellow, brown or grey. Eggs are oval, flattened and translucent. Larvae are small, less than 20 mm in length, with a soft cylindrical body and a hard head and shield on the first thoracic segment (prothoracic shield). Color of head, shield and body varies with the species. Pupae are spindle-shaped, less than 15 mm in length, and vary in color from green to brown.

A leafroller larva feeding on aspen leaf

M. Herbut

Moths emerge from the pupae sometime during the spring or summer, mate and lay their eggs singly or in flat clusters on the leaves, twigs or bark of host plants. Larvae hatch and either tie needles and leaves together or roll the edges of the leaves around a webbed shelter in which the larvae remain while feeding on the leaves. Feeding continues from 2 weeks to 2 months depending on the species. Pupation occurs in webbed leaves, in bark crevices or in the ground. The overwintering stage may be eggs, larvae or pupae. One or more generations occur during the year but one generation is the most common.

Food Hosts and Damage

Conifers and broadleaved trees and shrubs are hosts to leafrollers and leaftiers. Needles are tied together to form an unsightly cluster that gradually turns brown as the needles are mined or consumed. Trees heavily infested with leafroller larvae will appear somewhat defoliated because of the intensive leaf rolling. Larvae are protected from predators and chemical treatments inside the curled leaves and webbed nests. Nearby leaves may be chewed and eaten. Rarely will populations be great enough to damage or weaken host trees.

Control

Natural predators and parasites control these native insects, so mechanical and chemical controls are rarely necessary. Spraying the plants early in the season at budbreak with a systemic insecticide can reduce leafroller populations on nursery stock or ornamental trees. Removal and destruction of damaged leaves should be the only controls needed on small trees.

Lilac Leafminer
Gracillaria syringella (Fabricius)

Appearance and Life History

The lilac leafminer is an introduced insect that is found wherever lilac is grown. Adults are small (6 mm long), slender, greyish-brown moths with a wing-span of about 10 mm and irregular yellow patches on the forewings. Eggs are very tiny and transparent. Newly emerged larvae are green but become pale yellow with a yellow-brown head capsule as they mature. Full-grown larvae are flat and about 8 mm in length. Pupation takes place in a small whitish cocoon in the soil.

Lilac leafminers overwinter as pupae in the soil with the moths first appearing in late May or early June. After mating, the females lay about 100 eggs in batches of 5 or 10 along the midribs on the undersides of leaves. After 1 week, larvae hatch from the eggs, tunnel into the leaves and begin feeding on the internal tissue. After 3 weeks of feeding inside the leaves, the larvae, now about 5 mm long, come out of the leaf, roll the leaf downwards and feed on the inside of the roll. About 10 days later the mature larvae fall to the ground on silken threads and spin cocoons in which to pupate under ground debris. The second generation of moths emerge in early August and the larvae feed until about mid-September at which time they pupate to overwinter just below the soil surface. There are two generations per year on the prairies.

Food Hosts and Damage

Lilac leafminers feed primarily on the common lilac but other lilacs, ash and privet may also be attacked. The first sign of damage is a small linear mine on the under surface of the leaf but as the larvae grow, light green to brown blotches appear on the upper surface as they eat away the inner tissue of the leaf, leaving the upper and lower epidermis. The blotches gradually get larger, and eventually become brown and brittle. When the larvae leave the mines and roll up the leaves, they continue to feed and skeletonize the upper leaf surface within the curled leaves.

Lilac leaves damaged by lilac leafminer larvae

H. Philip

Most of the damage takes place in the lower part of the bush, especially the first-generation damage. Second-generation damage in August is usually higher on the plant. Damage is not considered fatal but it does give the bush an unhealthy and unsightly appearance.

Control

Hand picking mined and curled leaves and destroying them before the larvae drop to pupate will help reduce the population and improve the appearance of the bush. Spraying in late May and early June before the damage is noticed will help control egg-laying adults. A systemic insecticide applied to the foliage, stem or soil will also aid in reducing damage. Once the leaves are curled, insecticide treatments are not effective in preventing further damage.

Linden Looper
Erannis tiliaria tiliaria (Harris)

Appearance and Life History

The linden looper is an insect native to eastern Canada but found across the country. Males are light brown moths with a dark wavy band crossing the forewings and with pale brown hind wings. The wing-span reaches 42 mm. Females are wingless, 12 mm long, with two rows of black spots on the pale grey to brown body. Eggs are small, oval and orange-yellow in color. Larvae are bright yellow with ten wavy black lines running down the back. The bodies are elongate, up to 35 mm in length when mature, with orange-brown heads. Larvae move in the typical inchworm fashion, thus the name looper. Pupae are 12 mm long and glossy brown in color.

Linden looper larva feeding on aspen leaf

M. Herbut

Linden loopers overwinter as eggs under loose bark and in cracks and crevices of host trees. Eggs hatch in early May when buds burst. Larvae are solitary feeders and move about the tree, feeding on the leaves. They drop to lower leaves on a silken thread when

disturbed. After feeding about one month, the mature larvae drop to the ground, burrow up to 25 cm into the soil where they build earthen cells to pupate. Pupae remain in the soil during July and August. Adults emerge in the fall. Females crawl up the tree trunk while males fly at dusk, looking for the females. Eggs are laid singly or in small clusters before winter on the host tree. There is only one generation per year.

Food Hosts and Damage

Linden loopers feed on a wide variety of deciduous trees with American elm and Manitoba maple the preferred hosts in western Canada. Basswood, birch, poplar and other maples may also be attacked. Damage is caused by the conspicuous larvae which defoliate the host trees. Young loopers eat holes in the leaves while older larvae consume everything but the petiole. Total defoliation may occur if the population of loopers is high or if it occurs in conjunction with other spring defoliators. Damage is generally done fairly early in the season (June), and the trees may produce new foliage before the end of summer. Rarely do linden loopers cause death to host trees.

Control

Outbreaks do not generally last more than 1 or 2 years because of a number of natural predators and parasites including viral diseases. Controls are rarely necessary. Isolated shade trees may be protected by painting a wide band of sticky material around the trunk in late summer to trap the wingless females as they climb the tree to lay eggs. A contact insecticide on the bark may also work at this time. Spraying the foliage with a contact or stomach insecticide in late May or early June will control the feeding larvae.

Mourningcloak Butterfly

Nymphalis antiopa (Linnaeus)

Appearance and Life History

The mourningcloak butterfly is one of the most beautiful butterflies observed across the prairies. The wings are dark reddish-brown to black and bordered with a wide yellow stripe inside of which is a line of blue spots. Two yellow spots are found at the tip of each forewing. This butterfly has a wing-span of 60-80 mm. The undersides of the body and wings are mottled brown and black. The adults, after overwintering in sheltered locations such as hollow trees and under loose bark, are some of the first butterflies observed flying in the spring. The subcylindrical, ribbed, orange-brown eggs are laid in large masses or bands on twigs and branches of host plants in early summer.

The larvae, known as spiny elm caterpillars, are purplish-black with a row of orange spots down the back and short pale hairs and long forked spines distributed over the body. The five pairs of fleshy abdominal legs are orange-red in color. Mature larvae, 52-58 mm long, attach to twigs and molt into naked pupae or chrysalises which hang freely from the twigs. Adults emerge in August and September and go into winter hibernation in late September. One generation of mourningcloak butterflies is produced each year.

Mourningcloak butterfly larvae (spiny elm caterpillars) feeding on American elm leaves

M. Herbut

Food Hosts and Damage

Elms, willows, poplars and birches are the preferred hosts of the spiny elm caterpillar but other ornamental trees and shrubs may be attacked. Damage by young larvae is characterized by small holes and notches in the leaves. Mature caterpillars devour leaves completely. The young gregarious larvae feed in clusters and therefore tend to strip individual branches. Older larvae disperse over a tree with less noticeable damage. This insect is mainly a pest of ornamental trees in cities and parks.

Control

Picking or crushing or otherwise destroying egg bands is effective in reducing potential larval feeding damage. Pruning out or spraying individual infested branches provides effective control.

Northern Pitch Twig Moth

Petrova albicapitana (Busck)

Appearance and Life History

The northern pitch twig moth, also known as the pitch nodule maker, is a native insect found across Canada. Adult moths are small with a wing-span of about 20 mm and are variably and obscurely patterned with irregular bands of pale brown, shaded with dark grey. Eggs are small, 0.5 mm in diameter, yellow and flattened. Larvae are about 15 mm long, with a pale reddish body and a yellow-brown head and thoracic shield. Pupae are spindle-shaped with a black and reddish body.

Adults are active during June at which time eggs are laid singly on the terminal shoots of host trees. After hatching, the young larvae chew circular holes in the bark. These holes are lined with silk and pitch to form overwintering chambers or nodules. Feeding resumes in the spring and the nodules are enlarged. Larvae leave the first nodule in June and crawl to another twig or branch axil where they chew a second hole in the bark to create a larger silken chamber within a pitch-covered nodule. This nodule may be over 30 mm in diameter and serve as the second overwintering shelter. Larvae

pupate the next spring inside the nodule and adult moths exit through a hole cut by the larvae before they pupated. Larvae are solitary feeders and only one larva will be found in each nodule. The life cycle requires 24 months spread over three growing seasons.

Food Hosts and Damage

Jack pine is the preferred host of this insect, but it will also feed on Scots, lodgepole, red and mugho pine. Damage is caused by the larvae feeding under the bark and causing pitch galls or nodules. Girdling of young shoots may cause the needles to turn red and the terminal to become stunted, die and break off. Tree height may be reduced and branch shape may be distorted by larval feeding. The pitch nodules are conspicuous on the branches and damage is easily recognized. Open wounds left by the feeding larvae may permit the entrance of disease-causing organisms.

Northern pitch twig moth larval feeding damage to pine

H. Philip

Control

A number of natural parasites and predators generally keep this insect under control. The use of nonsusceptible pines and planting pines away from known infestations will help reduce damage. Hand picking and destroying the nodules on ornamental trees will aid in preventing damage to terminal shoots. Insecticides have not been used but may be useful if applied against adults to reduce egg laying or against larvae as they migrate to produce new nodules.

Sod Webworms
various species

Appearance and Life History

Sod webworms are the larvae of small brownish-white "snout moths" or grass moths that are recognized by their quick, jerky, erratic flights over lawns. Adults are small moths, 12-18 mm long, with wing-spans of 15-20 mm and a long "snout" projecting from the head. Wings are folded around the body at rest, giving them a cylindrical appearance. Wings and body may be marked with colorful, iridescent scales. The tiny oval eggs change from white to orange or red just before hatching. Larvae are creamy white to dark grey with light brown heads and bodies that may have dark spots and a few coarse hairs. Larvae may be up to 20 mm in length when full grown.

A sod webworm larva

H. Philip

Adults are active during July, August and September, flying over the grass at night, and scattering eggs over the lawn or placing them at the base of grass stems. They hide during the day in the grass, on tree trunks and in the shrubbery. Eggs hatch in about a week and larvae begin feeding on nearby host plants. Larvae feed at night, cutting off the grass blades and taking them to their protective silken cases in the grass crowns or soil. The young caterpillars overwinter in these silk cases. They resume activity in the spring and after feeding for a short time, spin grey silken cocoons to pupate just below soil level in early summer. Moths are noticed all summer due to the overlapping generations of the several species found on the prairies. The moths are readily attracted to lights at night.

Food Hosts and Damage

Sod webworms feed on various grass plants, more particularly bluegrass, timothy, other pasture and field grasses, and corn. Adults do not feed but larvae feed on grass leaves near the soil level. The first sign of damage is usually irregular brown patches of grass in late June and July. Continuous feeding combined with summer heat, drought and poor soil fertility may kill lawns and grasses.

The presence of large flocks of blackbirds, robins and other birds picking holes in the turf is an almost sure sign that webworms are present. Examination of the brown patches for evidence of chewed grass at or just above the soil surface may indicate sod webworm presence. Fresh clippings, green fecal pellets and the presence of webbing in the thatch will all indicate webworm presence.

Control

Damage by sod webworms can be reduced by maintaining vigorous plant growth. Healthy stands of turf can withstand more severe feeding than poorly maintained lawns. Sod webworms may be brought to the soil surface by applying a 1 percent pyrethrin drench to the lawn surface. The pyrethrins irritate the webworms, causing them to squirm to the surface in 5-10 minutes. If 15 or more larvae are found per square metre, insecticidal control is required to prevent further damage. The lawn should first be mowed and the clippings removed. Water the lawn thoroughly in the afternoon and apply the insecticide in the late afternoon or early evening. The lawn should then be watered lightly to wash the insecticide into the thatch and the soil.

Speckled Green Fruitworm
Orthosia hibisci (Guenée)

Appearance and Life History

The green speckled fruitworm is one of the most common of the group of moth larvae known as green fruitworms because of their green color and presence on fruit trees. Adult moths are 16 mm long and have a wing-span of about 40 mm. Forewings are greyish-pink, each marked near the middle with two purplish-grey spots outlined by a narrow pale border. Hind wings are lighter in color. Eggs are white with a greyish tinge and have numerous ridges coming from the centre. They are 0.8 mm in diameter and 0.5 mm in height. Larvae are greyish to green in color with several narrow white stripes along the top of the bodies and a slightly wider, more distinct white line along the side. Young larvae are 2-3 mm in length with a brown head and mature larvae may be 30-40 mm in length with a brown head. The brown pupae resemble cutworm pupae and are about 15-20 mm in length.

Speckled green fruitworm larvae feeding on wild rose leaves

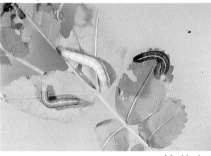

M. Herbut

Green fruitworms overwinter as pupae in the soil. Adults emerge and fly in the spring before fruit trees are in bloom. Females lay eggs on twigs and developing leaves. A female is capable of laying several hundred eggs, but normally deposits only one or two at each site. Newly hatched larvae feed on new leaves and flower buds and can often be found inside a rolled leaf or bud cluster. Older larvae damage flower clusters during bloom and continue to feed on developing fruit and leaves for 1-2 weeks after petal fall. They then drop to the ground, burrow 5-10 cm beneath the soil surface and pupate.

Food Hosts and Damage

Speckled green fruitworms prefer trembling aspen and willow but have been collected from a wide range of broadleafed trees. Damage to apple (*Malus* spp.) and cherry (*Prunus* spp.)

is common. Flowers and buds damaged by the larvae abort and premature fruit drop may occur. Fruit remaining may have deep corky scars and indentations. Leaf damage to poplar is slight.

Control

Control on aspen and willow is not usually required but may be necessary for fruit production on apple and cherry. Sprays applied against the larvae will reduce feeding damage and save the fruit.

Spring Cankerworm
Paleacrita vernata (Peck)

Appearance and Life History

The spring cankerworm is a minor pest of hardwoods on the eastern prairies. The moths emerge in the spring of the year and are therefore known as spring cankerworms as opposed to the fall cankerworms. Both species often occur together, feeding at the same time of the year and causing similar damage. Adults are also very similar in appearance. Males and females display sexual dimorphism with the males being winged and the females wingless. The forewings of the male moths are brownish-grey crossed with three dark lines and the hind wings are pale grey. The wing-span is about 35 mm. The wingless female is brownish to blackish with a black band down the middle of the back. Body length of female moths is about 10 mm. Eggs are oval in shape and white to yellow in color. The inchworm-type larvae vary from yellowish-green to reddish or blackish in color. The body may be marked with uneven black lines or a yellow stripe on the sides. Three abdominal dorsal segments are marked with a distinctive black X. Larvae have two pairs of prolegs and may reach 20-30 mm when full grown. The pupae are brown with a greenish tinge.

Spring cankerworms overwinter as pupae in earthen cells 5-25 cm below the soil surface. Adults emerge in the spring as soon as the ground thaws. Females crawl up the trunk, mate and begin laying eggs immediately. Each female lays up to 400 eggs in small clusters in bark

Spring cankerworm larva

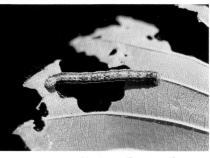

Northern Forestry Centre, Canadian Forestry Service, Edmonton

crevices or under bark scales on the trunk and branches of the trees. The eggs hatch and larvae begin feeding as soon as leaves appear and continue feeding through June. When feeding is complete, the larvae drop to the ground on silken threads and burrow into the soil where they build earthen pupal cells. They remain dormant in the pupal cells throughout the summer and autumn, pupate in the late fall or early spring and emerge as adults the following spring. There is one generation per year.

Food Hosts and Damage

The main host of the spring cankerworm is elm but many trees and shrubs may be attacked. Damage is similar to that caused by the fall cankerworm. Grey-brown moths flying around the base of trees early in the spring is the first indication that an attack by the spring cankerworm is in progress. Under close examination of the bark, wingless females may be found climbing up the trunk in search of an egg-laying site.

Control

Controls are similar to those for the fall cankerworm. Banding of the bark with a sticky substance for female control must be done early in the spring rather than in the fall.

Uglynest Caterpillar
Archips cerasivoranus (Fitch)

Appearance and Life History

Adults of the uglynest caterpillar are small, pale brown or yellow moths with a wing-span of 13-32 mm. There is usually

a prominent spot in the middle of the outer margin of the forewings as well as other irregular spots. The wings are wrapped about the abdomen or held roof-like over the back when the moth comes to rest. Eggs are laid in flattened clusters and covered by a whitish cement which turns black with age. Mature caterpillars are various shades of green, with a shining dark head and measure 19 mm in length. Body hairs are marked by a pale spot at their base forming broken stripes along the back and sides. Pupae are naked and brownish with sharp angular projections.

Uglynest caterpillars overwinter as eggs on the trunks and branches of host trees and shrubs. The eggs hatch in late spring at about the time leaves appear. All eggs hatch at the same time, usually within minutes of each other. Larvae crawl to the top of the host and immediately begin feeding and constructing a nest of webbing by tying together leaves, branches and fruit. This nest forms a compact hiding place for the larvae, protecting them from the weather and their natural enemies. When disturbed, the caterpillars wriggle backwards rapidly into the protective webbing. At maturity the larvae form silken cells inside the nest. Pupation takes place in these cells from late June to early September. The pupae wriggle to the outside of the nest where the adults emerge. Adult uglynest caterpillars are active from July through to September. Only one generation is produced annually.

Food Hosts and Damage

Uglynest caterpillars attack a variety of wild and cultivated fruit trees, ornamental trees and shrubs. The most common host is chokecherry. The first sign of an infestation is a few leaves joined together by a dense web. As the larvae grow and require more food, they pull additional leaves over and cover them with webbing until a dense tent-like nest is formed. In some cases the nest may envelope an entire shrub. Inside the nest, larvae skeletonize or devour most of the leaves and fruit. Chokecherry used for landscaping or as a part of shelterbelts is unsightly due to the "ugly nests" of webbing and defoliation of leaves. Infested shrubs are not usually killed.

Uglynest caterpillar nest in chokecherry

H. Philip

Control

Natural controls such as parasites, adverse weather conditons and predaceous insects and birds keep the uglynest caterpillar in check. Removal of the nests while still small provides adequate control if only a few shrubs are infested. New nests and larvae can be destroyed by burning or by pruning out the nests and disposing them in tied plastic bags.

Thysanoptera
Gladiolus Thrips
Thrips simplex (Morrison)

Appearance and Life History

The gladiolus thrips is one of the most serious insect pests attacking gladiolus. Adult female thrips are less than 2 mm long, shiny brownish-black with a conspicuous white band at the base of the forewings. Males are slightly smaller and the white band is not so conspicuous. The nymphs are about 1 mm long, and when full grown are a pale yellow color.

Gladiolus thrips overwinter on the corms. Under favorable conditions of humidity and temperature all stages may be present. In the spring, thrips are carried to the field on the corms and start a new cycle of infestation. Thrips may reproduce at temperatures of 15°C or above. Female thrips can lay up to 200 eggs, which are deposited in slits on the surfaces of the plant. Complete development from egg through adult may require as little as 10 days or as long as 45 days, depending on the temperature. Five or six generations may develop in a season. As the stems appear, the thrips crawl inside the leaf sheaths which offer excellent protection for their development. When the flower spikes develop, the insects crawl into the bud sheath where it is practically impossible to reach them with sprays or dusts. They thrive under hot, dry conditions, and can survive outdoors only where the ground does not freeze in winter.

Food Hosts and Damage

Gladiolus and iris are the preferred hosts but they have been reported on carnation, dahlia and lily. Damage occurs to the corms in storage and to the growing plant. Thrips feed by rasping the tender epidermal layers of leaves, corms and flowers, and feeding on the exuding plant juices. Infested corms become sticky, corky and roughened, and sometimes small rootlets may also be injured. Such damage usually results in

Gladiolus plant damaged by gladiolus thrips

H. Philip

retarded plant growth, poor flowers and smaller corms. Severely infested corms may fail to grow.

The growing plants show whitened or silvered areas on the leaves and flowers due to thrips feeding. Buds may be injured so severely that they never open normally; they turn brown and the bud sheaths dry out and become straw-colored. Leaves also brown and die. Shiny black fecal droplets may also be observed on plant surfaces.

Control

Gladiolus thrips can be controlled best by treatment of the corms prior to winter storage. Leaves and scales of the corms should be removed and the corms treated with an insecticide before being placed in storage. Corms should be stored in low temperatures of 2-4°C for at least 4 months to kill the thrips. Corms must not be frozen or they will be killed. A dichlorvos-impregnated strip hung among the stored corms in an area with little air circulation, or naphthalene flakes scattered among stored corms in containers covered with paper are effective methods for controlling thrips. Infested corms may be soaked in Lysol® solution for 6 hours before storage or before planting in the spring. Corms must be dried before storage. Foliar sprays can be used at the first signs of damage to the gladiolus plants.

Insect Pests of Households

Acari
Clover Mite
Bryobia praetiosa Koch

Appearance and Life History

The clover mite is often one of the first pests evident in the spring. Adults are tiny (smaller than a pin head), dark red, rusty-brown or olive-green in color and have eight legs. They are readily distinguished from other mites commonly found around homes or on plants by their very long front legs that extend forward from the body.

During the spring and summer months, clover mites are active at temperatures between 4°C and 24°C, feeding primarily on lawn grass by sucking sap from plants. In late fall, these mites migrate from their dwindling food supplies, sometimes in enormous numbers, to lay small, red, overwintering eggs in cracks and crevices on trees and buildings. Mites may be particularly numerous on the warmer, south-facing walls of buildings and may find their way indoors.

Overwintered eggs hatch the following April and May, earlier if the weather is favorable. Houses may once again be invaded by young mites in their search for suitable food plants.

Food Hosts and Damage

Clover mites are pests because of their presence in or on buildings, often in large numbers. They do not damage buildings, furnishings, humans or animals. They feed primarily on grasses, clover and other plants. They may also attack trees, shrubs and flowers, occasionally causing the foliage to gradually turn brown and wilt. They crawl into cracks around windows and doors or in foundation walls and under siding, shingles or shakes. This activity often leads many of the mites into homes where they can be seen on window sills, walls, tables, etc. When the mites are crushed or wiped up, they often leave a rusty or red stain on the cloth or surface. They are more likely to be a problem in areas of newly established gardens or lawns where there is a dense growth of succulent, well fertilized grass close to foundation walls.

Control

The first line of defense against clover mites is to seal off their entry points into the house. Caulking around windows, applying weather-strip around doors, and sealing foundation cracks all eliminate openings through which they may crawl. Carefully vacuuming infested areas in the house may remove mites without crushing them and leaving stains.

Mites do not readily cross loose, clean, cultivated soil; therefore a band 45-60 cm wide all around the house kept free of grass, will be a good deterrent. This strip may be planted with ornamentals but the soil around such plantings should be kept cultivated and free of grass, weeds and fallen leaves.

Pesticides may be used indoors directly on the mite or outdoors on the soil and foundation around the house.

Araneida
Shamrock Spider
Araneus trifolium (Hentz)

Appearance and Life History

The shamrock spider is commonly found in gardens during the late summer. Females measure 13-20 mm in length, and have a large round abdomen and long legs strongly marked with black rings at the joints. The head-thorax region is light colored with three wide, black stripes. The abdomen color varies from almost white with few markings to grey-olive, reddish-brown and purplish with variable markings. A three-lobed spot somewhat resembling a shamrock leaf is found on the midline near the base of the abdomen. Other spots and markings may also be present. Males measure only 4-8 mm in length, and have a slender white or yellow abdomen without distinctive markings.

Adult shamrock spider

M. Herbut

Shamrock spiders belong to a group of spiders called orb weavers that characteristically construct large orb webs in open areas between shrubs or in tall grass. The hub or center of the web is fairly open and meshed. The spider hides in a silken tent built in a folded leaf above and to one side of the orb. A trapline leads from the hub to the retreat where the spider waits with one foot on the trapline so that it feels when anything is caught. Adults mature in late summer, mate and lay eggs in round, delicate egg cases attached to leaves. Shamrock spiders overwinter in the egg stage. Only one generation is produced annually.

Food Hosts and Damage

The large, flexible, sticky webs are traps for many flying insects which are fed upon by predaceous shamrock spiders. These spiders are conspicuous because of their size and markings. They are not poisonous to humans and are actually beneficial to the gardener.

Control

Control of the shamrock spider is not necessary. Removing webs with a broom or hosing them down with a strong stream of water are the only controls recommended.

Chilopoda
Centipedes
various species

Appearance and Life History

Centipedes are common animals of the soil and leaf litter. They are long (2-4 cm), flattened creatures with from 15 to over 100 pairs of legs, one pair to each abdominal segment. The long, jointed legs provide for rapid movement. Long antennae are also characteristic. Most common species are greyish- to reddish-brown in color.

Adult centipede

Agriculture Canada, Ottawa

Centipedes appear to overwinter as adults, laying their eggs in dirt or damp places throughout summer. It may take a year to reach maturity and some have been reported to live 5 or 6 years. They are generally found in moist environments under boards, stones, piles of leaves, bark and crevices in damp soils. Because they lack the cuticular wax layer found in insects, they must live in areas where humidity is high. Very secretive in habit, they are usually nocturnal and run away when startled or exposed.

Food Hosts and Damage

Centipedes are predators, feeding on insects, spiders, earthworms and snails in the soil and litter. A pair of venom-bearing claws extending from under the head are used to stun, paralyze and often kill prey.

Since centipedes are carnivores, they do little or no damage to plants or buildings. Their presence outdoors is beneficial. Finding centipedes indoors, especially in basements, indicates the presence of other insects or food, possibly spiders or carpet beetles. Centipedes may bite people if handled. Occasionally, bites do break the skin and cause slight swelling and pain. However, these are no worse than a bee sting.

Control

Controls for centipedes are similar to those listed for millipedes and sowbugs. Elimination of moist hiding places such as piles of trash, rocks, boards, leaves, etc. is one of the first steps in control. Sealing cracks around windows and in basement walls will discourage their entry into the home.

Chemical controls can be applied if necessary.

Coleoptera
Black Carpet Beetle
Attagenus unicolor (Brahm)

Appearance and Life History

The black carpet beetle is found throughout the prairies. Adults are 2.8-5 mm long, dark brown or black in color with brown legs and oval in appearance. Eggs are small, fragile and pearly white. Newly hatched larvae are small and hairy. Mature larvae measure 9 mm long, are slender, reddish-brown in color and are characterized by a tuft of long hairs at the end of the abdomen. Each abdominal segment has a lateral tuft of stiff hairs and the whole body is covered with short, dark stiff hairs.

Adult beetles are usually noticed during the spring and summer months. They are attracted to sunlight and hence will be found on windows and on exposed surfaces outdoors during June and July. Mating and egg laying takes place 5-11 days after adult emergence. Females will

Adult black carpet beetle

M. Herbut

Black carpet beetle larva

M. Herbut

frequently enter houses to lay their eggs near a food source. Within 1 or 2 weeks, small, hairy larvae or grubs emerge and commence feeding. The larval period lasts 1-3 years, depending on temperature and food availability. Pupation usually occurs from April through June, with adults emerging about a month later.

Food Hosts and Damage

Adult carpet beetles feed on nectar and pollen of plants and have not been recorded as damaging household items. It is the larvae which are chiefly responsible for damage to wool carpets and furniture upholstery. Larvae will also feed on dried plant and animal products, including clothing. Infestations are detected by the appearance of damaged wool carpets and furniture, cast larval skins and adult beetles crawling around windows. Larvae are rarely noticed as they are repelled by light, and are found in undisturbed areas such as floor cracks, under carpets, behind baseboards, in collections of dust and lint in heating ducts, and in neglected trunks and cupboards.

Control

Preventing a carpet beetle infestation begins by protecting stored clothing and furnishings. Clothing should be cleaned before storing in clean, tightly sealed containers. Clothes can be protected by using moth crystals or by applying protective insecticides to the clothing. Rugs and furniture should be vacuumed and cleaned regularly. An infestation can be eliminated by thoroughly cleaning all garments, rugs and furniture. Hanging infested materials outdoors, either on a hot, sunny day in summer or on two or three successive sub-zero days in winter, will drive out or kill the larvae. Insecticides may also be used to control an infestation.

Ground Beetles
various species

Appearance and Life History

Several hundred species of ground beetles may be found on the prairies. Adult beetles are usually shiny, black or brown in color, and range in length from 5 to 20 mm. Most species have broad wing covers with many fine parallel lines or punctures. The body generally narrows towards the front. The hind legs characteristically have an elongated lobe (trochanter) at the base of the femur which readily distinguishes them from flour beetles and other pests. Their long legs enable them to run quickly to hiding spots. They have long antennae and large mandibles. Larvae are elongate, whitish, with brown heads.

A ground beetle

M. Herbut

Food Hosts and Damage

Their normal habitat is out of doors in the soil, under or amongst rocks, stones, debris and vegetation. The great majority of species are predaceous on other insects, though a few are vegetation feeders and several will scavenge. They are beneficial, but may invade buildings and become a nuisance by their presence if numerous.

Control

Ground beetles enter buildings for shelter through spaces between foundation and walls and through gaps around windows and doors. Attention to sealing these entry points will alleviate the problem, since breeding rarely takes place indoors and the population must replenish itself by more coming in from outside. Once inside they will not survive without food and water, so a thorough housecleaning and denial of access to water by covering drains and sealing water leaks is necessary. Such measures should be sufficient to keep these beetles down to an occasional invader. Find the beetles and dispose of them by hand. If necessary, an existing population can be controlled with household insect sprays.

Larder Beetle
Dermestes lardarius Linnaeus

Appearance and Life History

The larder beetle is a common household pest, especially during the spring. The adult beetle is 6-9 mm long, oval in shape, dark brown to black in color with a transverse, wide, pale yellow band at the base of the wing covers. The band has six dark spots. The undersurface of the body and the legs are covered with fine yellow hairs. Eggs are small, white, oval and rarely seen. The newly hatched larvae are white, gradually turning darker brown as they mature. Mature larvae measure 11-13 mm long, are covered with conspicuous hairs, and have a pair of short, curved spines projecting backwards from the top of the last body segment. The naked pupae are about the same size as the adult, but are soft-bodied without distinctive markings.

Adult larder beetle

M. Herbut

Adult larder beetles overwinter outdoors in bark crevices and enter houses in May and June to seek food on which to deposit eggs. If no food is available, eggs will be deposited in cracks and crevices in the vicinity of food stores. Each female lays 100-200 eggs. Eggs hatch in about 2 weeks, depending on the temperature, and the larvae commence feeding, reaching maturity in 40-50 days. When ready to pupate, larvae wander about in search of a convenient place to tunnel such as old rags, wood or similar compact material. The pupal period lasts 3-10 days, depending on environmental conditions. Beetles multiply until the source of food is consumed, after which adults and larvae move about in search of further nourishment. There are two generations produced annually.

Food Hosts and Damage

Adult and larval larder beetles feed on a variety of materials of animal origin such as feathers, animal skins, hair, ham, bacon, dried and processed meats, decayed meat, cheeses, dead insects, and wool. In recent years they have been found in increasing numbers in dry pet foods containing a mixture of cereal and animal products. The presence of dead mice between walls of the house and accumulations of dead insects in lamps, between walls or at attic windows may also support large numbers of larder beetles. The larvae may also damage wood products by burrowing into them to pupate. They will also tunnel to pupate in ham or bacon which are recorded as preferred foods. Infestations are characterized by the presence of adult beetles and larvae moving about a house in search of food and oviposition sites in the case of the adults, and food and pupation sites in the case of the larvae.

Control

The beetles are easily seen and may be caught by hand and killed. This is satisfactory if they are not too abundant. Cheese may be used as a bait to attract and capture beetles. Eliminating food sources such as cleaning mouse traps, removing old bird nests and cleaning light fixtures of dead insects is probably the best control for larder beetles. Sealing baseboards and other openings in walls and attics will help reduce migration of larvae and beetles into rooms. A residual household spray can help if applied as a spot treatment along baseboards and other areas where beetles are noticed. Smoked and cured meats should be wrapped and placed in cold storage. Insects in dry pet foods may be killed by heating to 50°C for 30-45 minutes or by freezing for several days.

Merchant Grain Beetle
Oryzaephilus mercator (Fauvel)

Sawtoothed Grain Beetle
Oryzaephilus surinamensis (Linnaeus)

Appearance and Life History

The merchant grain beetle and the sawtoothed grain beetle are very similar in life history, appearance and habits. Adults of both species are dark brown, slender, about 3 mm long, and are characterized by six saw-like projections on each side of the thorax. The pronotum has three low longitudinal ridges and the antennae end with a compact club. Eggs are extremely small, white and slender. Larvae are 3 mm long, white to pale yellow with darker plates on the top of the thorax and abdomen. Mouthparts face directly forward from a flattened head and the antennae are about as long as the head. The last segment of the abdomen is tapered without large dorsal projections. Pupae are white with short wing covers and may be found in cocoon-like coverings made of grain fragments cemented together.

Adult merchant grain beetle

M. Herbut

Adult specimens of these two species may be separated by examining the projection immediately behind the eyes. Merchant grain beetles have a narrow pointed projection behind the eye, whereas in sawtoothed grain beetles the projection is broad and rounded.

The female lays a total of about 400 eggs singly or in small groups in sheltered, undisturbed locations. Eggs hatch in 3-5 days in summer and in 8-17 days in colder weather. Larvae feed for 2-10 weeks before pupating. Adults emerge 6-21 days later, mate and begin laying eggs in about a week. Adults generally live about 6 months but have been known to live up to 3 years. There may be 4-6 overlapping generations a year depending on the temperature and humidity. All stages of these pests may be found during the year.

Food Hosts and Damage

The merchant grain beetle is the most common pest of household stored products on the prairies. The flattened shape of the adults enables them to enter packages of food that are apparently tightly sealed. Foods attacked include almost all plant products used for human consumption. Only damaged stored seeds are attacked by adults, hence they are often found with other stored grain insect pests which create the initial damage. The larvae are generally found in packages of food, whereas the adults feed not only on packaged foods but also on food bits and debris collected in cupboards, drawers and along baseboards in kitchens. Adults are very active and move about freely in search of food.

Sawtoothed grain beetles feed on dry grain, destroying the germ in the kernels. They may also occur in milled grain products such as cereals, bread and macaroni. They can overwinter outside as adults whereas the merchant grain beetles can only survive in heated buildings.

Control

Prevention of beetle infestations is the best approach to control of these pests. Indoors, the merchant grain beetle often enters in purchased cereal products. Examination of food items, purchases in small quantities, storage in sealed containers and removal of crumbs and spilled flour products will aid in preventing infestations. Once found, a thorough cleaning of the cupboards and removal of contaminated food items is required to rid the premises of this pest. A residual insecticide may be used if necessary.

Sawtoothed grain beetles in grain bins or ground grain may be controlled using similar techniques as described for the rusty grain beetle.

Strawberry Root Weevil
Otiorhynchus ovatus (Linnaeus)

Appearance and Life History

The strawberry root weevil is a humpbacked dark brown to black, hard-bodied insect, 6 mm long with a short, blunt snout and elbowed antennae. Adults are unable to fly. Both adults and larvae (grubs) overwinter under soil trash and around the base of food plant hosts. Adults emerge from their overwintering sites in late spring and early summer and migrate in search of food hosts around which to lay eggs. They also migrate in late summer in search of egg-laying and overwintering sites.

Male weevils are unknown. Females are capable of laying eggs without mating. Each female lays 200-300 small, white, spherical eggs in cracks in the soil near the crowns of plants. The small, whitish, legless larvae hatch in 2-3 weeks and immediately begin feeding on the roots of host plants. Mature or fully grown grubs

Adult strawberry root weevils

H. Philip

measure up to 12 mm in length, have pale brown heads, and hold their bodies in a more or less curled position. Larvae which hatch from eggs laid early in the summer are transformed into adults in the fall. Larvae which hatch from eggs laid in late summer overwinter in the soil and are transformed into adults the following spring. Females die after egg laying.

Food Hosts and Damage

Adult strawberry root weevils do little noticeable damage except for light feeding on the leaves of strawberry, clover, various grasses and assorted weeds. Leaf feeding is characterized by small notches along the leaf margins. Adults will feed on ripening strawberries, causing small holes in the fruit. Most economic damage is caused by the larvae feeding on the roots of strawberry, raspberry, various clovers, grasses (especially timothy) and nursery evergreens. When abundant, larvae can seriously damage strawberry plants which will appear stunted, the leaves closely bunched and dark colored or dying. Seedling evergreens are often weakened or killed by larval feeding damage to the roots.

Most complaints about strawberry root weevils occur because of their abundance in early and late summer in and around homes. When migrating, the weevils go through houses rather than around them, gaining access through spaces beneath doors, around basement windows and between the house and foundation. When abundant, they can be found crawling on floors, walls, ceilings, in sinks and bathtubs and on bedding and clothing. They do not feed and are quite harmless in homes but

their presence often causes alarm and concern.

Control

Outdoors - Rotate strawberries with non-susceptible crops. Use insecticides as recommended.

Indoors - Since the insects are always on the move during migration, several insecticide applications would be necessary to completely stop weevils entering homes. This is not practical nor acceptable from a potential health hazard standpoint. Therefore the best control procedure is to prevent weevils from entering buildings by ensuring there are no cracks or spaces beneath doors, door frames, basement windows, etc. Seal any crack with caulking compound or other suitable material, and install weather-strip where required.

If weevils do manage to enter, use a broom or vacuum cleaner to collect and remove them from the building rather than spraying them. A residual insecticide may be used around the house and foundation on the lawn and ground for 50-60 cm from the building foundation.

Dermaptera
European Earwig
Forficula auricularia Linnaeus

Appearance and Life History

The European earwig is an insect introduced into North America and can now be found in most inhabited regions. Earwigs are shiny, dark, reddish-brown insects about 15-20 mm long with pale yellowish-brown wing coverings and legs. They can be easily recognized by the pincher-like "forceps" on the tip of the abdomen. The larger male has strongly curved forceps while those of the smaller female are almost straight. Earwigs have long and flexible antennae and chewing mouth parts. The flying wings are small and fan-shaped, folding under the short, leathery wing covers at rest. Their bodies are long, slender and flattened, allowing them to crawl into narrow crevices and hiding places.

Adult European earwigs

M. Herbut

Earwigs overwinter as adults in nests built 3-5 cm in the soil. Females begin egg laying in late winter or early spring and produce a mass of 40-60 eggs. The female tends the eggs, turning and licking them constantly until hatching. She will guard and care for the early nymphs until they can fend for themselves, at which time both the female and nymphs abandon the nest. Since they undergo simple metamorphosis, the nymphs look similar to the adults but are much smaller. Only one generation is produced per year.

Earwigs are nocturnal in habit and thrive in cool, moist situations. By day they hide in moist, secluded places under stones, boards, bark, leaf and petal whorls, hollow stalks or woodpiles. By night they forage freely and may climb up on posts and launch themselves into the air. They are generally considered weak fliers.

Food Hosts and Damage

Earwigs are mainly omnivorous scavengers, feeding on living and decaying plant or animal material. They enjoy pollen, lichen, ripe fruit and succulent plant material. They will attack aphids and other live insects, grasping them in their forceps and twisting their bodies around to feed on the prey. They have been introduced as aphid predators in some apple orchards.

Earwigs are not generally considered destructive and their pest status is more a nuisance indoors with some plant problems outdoors. Their repulsive appearance, unpleasant odor and simple presence makes them a pest indoors. Easily transported indoors on cut flowers, vegetables, boxes and folded newspapers, their sudden appearance in kitchen refuse can be startling to some.

Outdoors, damage is generally confined to succulent plant material or flowers. Their feeding and tunneling into plant material can be a problem in ripe fruit such as peaches or strawberries and leafy vegetables such as lettuce. Pollen, a protein source, is often taken from flowers such as dahlias, zinnias, hollyhocks, and sweet williams.

Control

Since earwigs prefer moist and dark places, sanitation or removal of congregation sites is the major way to reduce populations. Removal of leaf litter and debris near foundation reduces overwintering sites and removes their protection from sunlight and desiccation. Doors, windows and vent pipes must fit tightly to prevent their entry into houses. A residual spray may be used along the foundation to prevent invasion. Mid-summer spraying around hiding places may aid in control in the garden. However, because of their activity, earwigs quickly re-infest sprayed gardens. The use of poisoned bran baits has had some success as a control in some areas.

Dictyoptera
Cockroaches
various species

Appearance and Life History

Adult cockroaches are broad, flattened, oval insects with long, slender antennae and long, narrow, spiny legs. They vary in length from 18 to 40 mm and are normally dark brown, reddish-brown, light brown or black in color. Adults of most species have wings; all possess chewing mouthparts. Eggs are laid in leathery capsules or egg cases called oothecae which resemble brown beans or small shredded wheat biscuits. A single egg case is carried protruding from the end of the female abdomen. One egg case may contain 14-30 eggs. Newly hatched nymphs are identical to their parents in appearance except they are much smaller and lack wings.

Cockroaches grow by gradual metamorphosis, having three stages of growth: egg, nymph, and adult. Females deposit egg cases in out-of-the-way places near food supplies. Egg cases are fairly characteristic and can be used as an identification aid. Nymphs may molt 5-13 times before reaching adulthood depending upon the species and local conditions. One to three generations can be produced annually. Generations overlap, so all stages may be found at any time of the year.

Roaches generally hide during the daytime, emerging only in darkness from cracks and crevices in walls, floors, etc. in search of food. If disturbed, they run rapidly for shelter and disappear through openings to their hiding places.

Cockroaches are not native insects but are regularly imported in contaminated food shipments.

The German cockroach, *Blattella germanica* (Linnaeus), is the most common cockroach found on the prairies. Adults are about 14 mm long and light brown with two dark brown stripes running lengthwise on the back. They prefer warm moist locations such as kitchens and bathrooms but can be found in most rooms in an infested building.

The American cockroach, *Periplaneta americana* (Linnaeus), is the largest cockroach found on the prairies. Adults measure up to 40 mm in length, are reddish-brown to dark brown with a light yellow or tan band around the edge of the shield behind the head. The large leathery wings fold flat over the abdomen. They prefer warm, damp basements, furnace rooms, storage rooms and sewers in commercial and industrial districts.

The oriental cockroach, *Blatta orientalis* (Linnaeus), is about 25 mm long when mature, dark brown or black with very short wings not extending to the tip of the abdomen. This slow-moving, strong-smelling species prefers damp, cool, dark areas near sewer drains and basements.

Female American (upper) and German (lower) cockroaches with egg cases

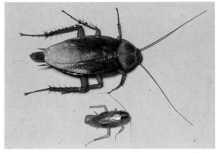

H. Philip

The brownbanded cockroach, *Supella longipalpa* (Fabricius), is light brown, about 13 mm long with a light yellow or pale brown band across the base of the wings and another broken band a third of the distance from the base. This species prefers warm, dry sites and may be found hiding behind objects hung on walls in homes and offices.

Adult brownbanded cockroach

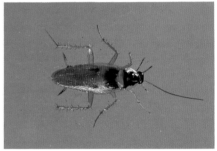

M. Herbut

The Australian cockroach, *Periplaneta australasiae* (Fabricius), is reddish-brown to dark brown with yellow markings on the thorax and yellow streaks on the base of the wing covers. Adults measure about 25 mm in length and prefer very warm, damp locations.

Adult Australian cockroach

M. Herbut

Food Hosts and Damage

Cockroaches feed on a wide variety of materials but prefer vegetables, cereals and meat products. They often feed on pastries, chocolate, milk products, beverages, cooked potatoes, glue, book bindings, wall paper, animal food, fresh or dried blood, excrement, dead animals and leather products. They are most abundant where food is stored, and near moist or humid locations such as floor drains, sinks and water pipes. Once established, cockroaches can be a serious nuisance in homes, restaurants, food processing plants, hotels, food warehouses, ships and any other place which offers food, warmth and shelter.

Cockroaches may carry organisms that cause food poisoning, dysentery and diarrhea. Most cockroaches produce a secretion that has a repulsive odor and affects the flavor of food. This characteristic odor can be detected in infested areas. They can cause allergic reactions in some people when roach "allergen" is ingested with contaminated food or inhaled when dried fecal particles and powder from the ground-up bodies of dead roaches are mixed with house dust. The sight of cockroaches can cause considerable psychological or emotional distress in some individuals. Cockroaches usually do not bite, but their leg spines may scratch.

Control

Inspection of incoming merchandise such as food boxes, beverage containers, appliances, furniture and clothing will help prevent cockroach entry into buildings. Cleanup of food, dishes and spilled garbage removes any food sources that may be attractive to roaches. Removal of breeding sites such as paper boxes, piles of paper bags, newspapers and soiled clothing will also discourage roach survival. Sticky traps placed along walls may be used to reduce and monitor a cockroach infestation. Residual insecticides applied where roaches are located or seen to frequent combined with good sanitation practices will eliminate infestations.

Diplopoda
Millipedes
various species

Appearance and Life History

Millipedes or "thousand-legged worms" are slender, wormlike creatures adapted to life in leaf litter of the temperate and tropical broadleafed forests. Their bodies are rounded as opposed to the flattened form of centipedes and they have two pairs of short legs on most segments while centipedes have only one pair. They are rather slow-moving in comparison to centipedes. Millipedes also have fairly short antennae, a small eye patch located on each side of the head and chewing mouth parts. Millipedes may be brownish, dark grey to reddish-black in color and are usually 25-38 mm in length.

Adult millipede

Agriculture Canada, Ottawa

Millipedes are nocturnal in habit, hiding by day beneath logs, stones and leaves and in cracks and crevices. They lack the waxy cuticle of insects and must therefore live in humid or damp situations to prevent water loss. When disturbed, they lie on their sides curled up in a tight ball. This defensive position protects their soft underbody and exposes the hard shell of their backs to predators. Some millipedes may release a stinking secretion from glands between the body segments. This secretion may be toxic to small animals and may cause irritation to human skin.

Millipedes overwinter in the soil near building foundations or under debris near tree trunks. Mating occurs in the fall and eggs are laid in clusters in the soil, sometimes in a capsule prepared for them by a female during the spring.

Hatching occurs in the spring and larvae pass through 7-10 molts over a 2-year period before reaching sexual maturity. The number of legs and segments generally increases with each molt, and the number of molts, legs and segments vary among species. Adults may live for 4 or 5 years.

Millipedes swarm in the fall, collecting along walls, trees and often invading houses. This swarming behavior may be for mating, finding an overwintering site, or perhaps their environment became too dry or too wet for survival and a new home is sought. They get into buildings through any small opening - a crack in the foundation, around a window or door frame, screen, sash, or under a poorly fitted door.

Food Hosts and Damage

Millipedes are generally considered harmless and usually beneficial. Most millipedes are humus feeders and play an important role in forest-floor ecology, breaking down dead plant material digestively and loosening the humus mechanically by burrowing. Damp and decaying wood and vegetable matter is by far their preferred food but they have been reported feeding on tender roots and green leaves on the ground. They have been reported damaging turnips, potatoes and the fruits of strawberries. Their entry into houses is generally considered accidental but they may feed on tubers or vegetables stored in cellars and basements. They do not survive well under the warm dry conditions normally found indoors. Some introduced species occasionally are blamed for damage to greenhouse crops.

The greenhouse millipede, *Oxidus gracilis* (Koch), is more flattened than other millipedes, has 30 or 31 pairs of legs and measures 20 mm in length. This millipede varies in color from creamy white larvae to deep chestnut-brown or black adults. It is often found in greenhouse situations, feeding on decaying leaf litter and other organic matter. Plant damage by this species is questionable because under research tests it could not be induced to feed on tender living plants.

Greenhouse millipedes in potted houseplant soil

M. Herbut

Control

The reduction of organic matter and drying out of soils reduces millipede numbers. Steam sterilization of soils used for potted plants greatly reduces their chance of survival. The removal of decaying wood around new houses also helps to reduce their numbers. In basements, they can be easily swept up with a broom and removed. Spraying around foundations during swarming may be necessary to prevent invasion of millipedes into a building. The use of a soil drench for infested potted plants may be required in some circumstances.

Grylloptera
Camel Crickets
Ceuthophilus spp.

Appearance and Life History

Camel or cave crickets are not true crickets but wingless long-horned grasshoppers. They have a fat, rounded body, arched back, jumping hind legs and long, slender antennae. Adults measure 25-30 mm in length and are light yellow to brown in color. Nymphs closely resemble the adults.

Adult camel (cave) cricket

H. Philip

Female camel crickets lay their eggs in small depressions in soil, rotten wood or rodent nests. They overwinter in the egg stage and only one generation is completed each year.

Food Hosts and Damage

Camel crickets are not of economic importance, however, their presence in basements of houses alarms many home owners. Outdoors, camel crickets are nocturnal, secretive insects feeding on a variety of organic material. They only enter cool, dark and damp basements towards the end of the summer during persistent hot dry weather. Disturbing the soil and hiding places around buildings will also force them indoors.

Once inside a dwelling, they will feed on any available cheese, vegetables, jams, fruits, fresh and dried meat, sugar and dead insects. They will not feed on cereal products such as bread or flour.

Control

They are slow-moving, thus they are easily captured and disposed of outdoors. Insecticides around windows and cracks in the basement may be applied if the crickets are numerous. Installing weather-strip around doors and windows, and caulking cracks between house and foundation will prevent entry into dwellings.

Heteroptera
Boxelder Bug
Leptocoris trivittatus (Say)

Appearance and Life History

The boxelder bug, a native North American insect, is an occasional household pest each fall. Boxelder bugs are rather flat, elongated insects 12-14 mm in length. They are dark grey to black with three distinct red lines behind the head and red lines on the wings. The abdomen under the wings is red but the legs, antennae and head are black. Eggs are ovoid and bright red in color. Nymphs resemble adults, are bright red with dark heads, gradually becoming marked with black and developing wings as they mature.

Boxelder bug nymph (left) and adult (right)

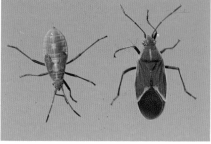

M. Herbut

Adults come out of their overwintering sites in early spring and mate. Females lay their eggs singly or in groups of up to 12 on the bark of the host tree, on the soil surface, and on grass, fences, sides of buildings and miscellaneous objects. Eggs hatch in about 2 weeks. Nymphs feed by piercing plant tissue with their pointed mouthparts and sucking the juices from leaves, fruits or soft seeds. Nymphs pass through five growth stages until mid-summer when they develop into adults. Adults overwinter in buildings, under debris and under loose bark of trees. Dispersal is primarily by flight in the spring and fall. There is only one generation per year.

Food Host and Damage

The primary host plant is the female (seed-bearing) Manitoba maple tree, also known as boxelder. Adults and nymphs also feed on ash and other maple trees, and on the fruits, flowers and foliage of apple, cherry, grape, peach, plum, raspberry, strawberry and the foliage of potato. Fruits may be unmarketable due to the feeding punctures which cause deformations. Bugs feeding on foliage cause the leaves to curl up and appear blistered. Flowers will wilt and drop off if severely attacked. Plants are seldom damaged seriously enough to justify control.

Boxelder bugs are more important as a nuisance during the fall and on warm days in winter when they swarm into houses or congregate in great numbers on trunks of trees, porches, walls and walks. They enter homes through small cracks in or above the foundation. Inside the home they may cause stains on curtains and walls. They do not feed on anything in the home nor bite humans or pets.

Control

Removal of Manitoba maple trees from the premises may help reduce boxelder bug numbers. However, they can fly far enough to invade a home well away from host plants. Wide-area removal of Manitoba maples is not suggested, as the shade value of trees is more important than the bug nuisance. When planting shade trees, give preference to species other than Manitoba maple. In summer, caulk all openings around windows, doors, in walls, foundations and siding. The best method of control after the insects enter the home is to vacuum as they appear, and then destroy them. If only occasional specimens are observed, use a fly swatter. Sprays may be directed against them inside or around the outside of the home.

Hymenoptera
Carpenter Ants

Camponotus spp.

Appearance and Life History

Carpenter ants are the largest and most conspicuous ants on the prairies.They can be distinguished from other ants by the evenly rounded or humpbacked thorax while other ants have a notch or depression on the thorax when viewed from the side. The brownish or blackish workers are from 6 to 13 mm in length while the winged queens may be over 25 mm long.

Carpenter ants are social insects living in nests with all stages of their life cycle present. The translucent white eggs are elongate, elliptical and about 0.5 mm long. Eggs incubate for 16-27 days. The larvae are soft, legless, translucent yellowish-white and vary in size depending upon the ultimate adult form (male, female or worker). Larvae develop for 8-14 days depending upon the food supply and form. Creamy white pupae are generally encased in elliptical, papery, light brown cocoons. The pupal stage may last from 14 to 92 days. The normal time required to develop from egg

to adult is about 60 days. Adults display polymorphism, that is, there are a number of different castes of shapes and sizes depending upon the sex and the work the adults will do.

Carpenter ant workers excavating a spruce board

M. Herbut

Carpenter ants get their common name from the fact that their nests are made out of wood. They live primarily outdoors with their nests built in stumps, logs fence posts, telephone poles and butts of logs. Carpenter ants remove softened (rotten) wood to make galleries in which to rear young and hold reproductive males and females. They do not eat wood but remove wood fibers and sawdust from their tunnels using their large strong mandibles. Their principal food is honeydew excreted by aphids, however, they also prey on a variety of insects. Foraging adults come out at night and may be scattered up to 90 metres from the nest. They do not lay trails as other ants do.

Winged males and females emerge from mature colonies in the spring and early summer to swarm, mate and begin new colonies. The males die shortly after mating but the females (queens) may return to old colonies or begin new ones. They will chew off their wings, tend the eggs and feed the newly hatched larvae until they become adults and can forage for food for the colony. Queens may live up to 15 years.

Food Hosts and Damage

Carpenter ants become pests when they move into buildings and build their nests in structural timbers, hollow panelled doors, window frames and wooden porches. Indoors, carpenter ants prefer to nest in moist decaying wood,

however, they will tunnel through dry timbers. Their presence in buildings may be particularly annoying because they may bite with their powerful jaws. Finding them indoors at anytime of the year indicates the presence of a colony nearby. Swarms of large, black, winged ants in the house, usually in spring, also indicates an infestation.

Finding the nest usually is the most difficult step in control. Piles of sawdust at the base of posts, along sills, baseboards or elsewhere are indicators of activity. However, the sawdust may not always be evident as it accumulates in walls, under floors or in attics. Following the workers is not helpful as they do not usually follow a path or trail marking. Hearing a distinct dry, rustling sound may help locate a nest. The sound will increase in volume if the ants are disturbed by pounding near their nest. A knife blade inserted at this point will readily penetrate infested wood.

Control

Several precautions can be taken to prevent carpenter ants from infesting a building. A tightly constructed house with concrete foundations, good clearance and a full basement is least subject to infestation. General sanitation of the building site by the removal of stumps, logs and wood debris will be a good start. Storage of firewood supplies away from the house and off the ground will help prevent bringing the ants into the house. Wood in contact with the soil should be treated. Any ant colonies within 100 metres of the house should be destroyed. This may be done with residual chemicals or baited food sprayed on or into the colony or its entrances. Insecticidal baits are especially effective because the treated food particles are taken back to the nest and fed to the queen and larvae. Controls that do not kill the queen usually are ineffective. The use of residual sprays along baseboards or where foraging ants are regularly observed will also help in reducing the colony.

Isopoda
Sowbugs and Pillbugs

Appearance and Life History

Sowbugs and pillbugs are the only land-dwelling crustaceans. These European invaders belong to the same class of animals as crabs, shrimp and lobsters. Their flattened, oval and greyish appearance reminds one of armadillos, especially with the series of thickened tough plates covering their backs. They are wingless inhabitants of the moist litter in the garden and greenhouse.

Adult sowbug

Lloyd Harris,
Saskatchewan Department of Agriculture

Sowbugs and pillbugs may be up to 15 mm in length but most are slightly smaller. They have seven pairs of legs as opposed to three pairs in the insects. They do have antennae, well developed compound eyes and generalized chewing mouth parts. Sowbugs differ from pillbugs in that they cannot roll into a ball like pillbugs when disturbed. This characteristic affords good protection from other soil-inhabiting predators such as spiders and centipedes. Sowbugs also have two prominent tail-like appendages which pillbugs do not have. Neither sowbugs nor pillbugs have the waxy coating of insects so they must remain in moist situations to prevent water loss.

Both sowbugs and pillbugs overwinter as adults and mate in the spring. The female may carry from 20 to 200 eggs in a pouch on the underside of the body. Eggs hatch in 3-7 weeks with the young remaining in the pouch for another 6 weeks. Once the young leave the pouch, they must feed on their own and will

reach maturity in about one year. Adults may live up to 3 years.

Sowbugs and pillbugs prefer living in an area high in organic matter and high humidity. These creatures are active at night when the temperatures are lower and the humidity is higher. Although outdoor pests, they may occasionally be found in greenhouses or in damp basements. The leaf litter under shrubs and around buildings makes an ideal habitat.

Food Hosts and Damage

Sowbugs and pillbugs are not serious pests and their damage is generally slight. Their habit of feeding on decaying plant and animal material aids in soil building. They may occasionally injure tender shoots and young plants in the home and in the greenhouse. Moist wood in a basement may rot more rapidly when sowbugs are present.

Control

Cultural controls require the elimination and drying of moist environments and shelters. Removal of excessive leaves, mulch and grass clippings from around buildings will certainly aid in drying up the area. Removal of boxes, boards and flower pots where they may hide will also be of help. Basements and crawl spaces must be kept free of rotting wood, decaying vegetables, soils and other materials attractive to them. Airing out basements where they have become established is also necessary. Foundations, doors and windows should be properly sealed to prevent entry of the sowbugs. If found in other parts of a home, sweeping with a broom or vacuuming may be all the control necessary.

Chemical sprays around foundation or gardens may sometimes be necessary.

Lepidoptera
Indian Meal Moth
Plodia interpunctella (Hübner)

Appearance and Life History

The Indian meal moth is found across Canada in granaries, homes and stores. Moths are about 10-12 mm long with a wing-span of 2 cm. They are distinct from other moth pests in homes by the coloration of their forewings – light grey on the inner third and reddish-brown with a copper luster on the outer two-thirds. The two areas are separated by a dark brown line. The minute eggs are white. Larvae have dirty white bodies, often tinted pink or green depending on the food consumed, and light brown heads and prothoracic shields. Mature larvae measure 10-12 mm in length. The light brown pupae are found in silken cocoons spun by the larvae.

Adult Indian meal moths

H. Philip

Each female moth lays from 100 to 300 eggs singly or in clusters of 12-30 eggs in or near a food source suitable for larval feeding. Eggs hatch in about a week and the young larvae disperse into the food material, constructing tunnels of silk and frass. They generally stay near the surface of the food and spin a silken web, underneath which the larvae are protected from changes in their environment and insect parasites. When full grown, the larvae pupate in cocoons generally found on the surface of the food, but some are found on braces, rafters or beneath floors. Adults begin emerging in about 1-2 weeks, mate and the life cycle is repeated.

Adults are most active at night, hiding during the day in concealed locations and, if disturbed, flying in irratic patterns. Depending on temperature and food availability, larval development may take anywhere from 2 weeks to 2 years to complete. Under favorable conditions, a life cycle can be completed in 4 or 5 weeks, and there may be as many as six generations per year.

Food Hosts and Damage

Indian meal moth larvae feed on grains, nuts, peas, beans, dried fruits such as raisins and other dried foodstuffs in the home, store and granary. Broken kernels of grain and the germs of grain kernels may be attacked as larvae have difficulty penetrating undamaged grain. Infested food material is usually webbed together with silk containing larval frass making it unsightly and inedible. Larvae often wander about kitchens and nearby rooms in search of suitable places to spin their silken cocoons and pupate.

Another closely related pest with similar food preferences, life cycle and behavior, is the Mediterranean flour moth, *Anagasta kuehniella* (Zeller). This occasional pest is smaller, 9-10 mm long, and has grey forewings with diagonal zigzag black lines.

Adult Mediterranean flour moth and forewing

M. Herbut

Adult meal moth

M. Herbut

The meal moth, *Pyralis farinalis* Linnaeus, is slightly larger than the Indian meal moth. The forewings are light brown with dark brown patches at the bases and tips with two wavy, transverse white lines. Meal moths prefer damp or spoiled grain, bran or meal.

Control

Sanitation of storage areas and spilled foods will reduce the likelihood of an infestation. Storage of flour or other host materials at low temperatures or in sealed containers is recommended, especially if the food materials are stored for more than a month. Infested and damaged food should be discarded as no satisfactory way of separating the insects from the food products, flour or meal have been found. Infested grain and flour can be fumigated.

Pseudoscorpionida
Pseudoscorpions
various species

Appearance and Life History

Pseudoscorpions, or false scorpions, are common predators in ground litter. Although quite common, pseudoscorpions are usually overlooked because of their small size (4-10 mm long) and secretive behavior hiding in cracks and crevices. They look very similar to scorpions with their enlarged pinchers but their bodies are flat and rounded posteriorly and they lack the stinging tail of scorpions.

They are usually golden or reddish-brown in color. The large pinchers on the forelegs are the most conspicuous diagnostic feature of these animals. The pinchers not only catch and hold the prey but also poison the prey from venom injected through ducts in the moveable fingers.

Silk is secreted to build a protective cocoon for molting, egg laying, rearing of young and overwintering. An adult pseudoscorpion may live 3-4 years. Both males and females occur but fertilization is usually by an external sperm case placed on a stalk.

Adult pseudoscorpion

M. Herbut

Pseudoscorpions also have the habit of hitch-hiking from one place to another. They hang onto other larger insects or daddy-long-legs with their pinchers. Since they are not parasitic, this behavior by hungry, mature, mated females may turn out to be a grab by mistake or a free ride to a new neighborhood.

Food Hosts and Damage

Pseudoscorpions are considered harmless as pests of humans. Outdoors they may be found in leaf mold, in nests of birds and mammals, under stones and beneath bark. Their main food is springtails, mites and small flies while other litter-inhabiting insects such as small beetles, ants and larvae may be eaten. They are occasionally found in homes, especially in bathrooms, probably entering on clothing.

Control

They do no harm and may be swept up or killed by crushing.

Psocoptera
Psocids
various species

Appearance and Life History

Psocids, (pronouned so'sids), also known as booklice or barklice, are small insects, 1.5-5 mm long, with soft, stout bodies that may have delicate membranous wings. They have chewing mouthparts and long 13- to 50-segmented antennae. They may be pale white, greyish-white, or yellowish in appearance. Development is by gradual metamorphosis (the young look like the adults) and individuals with four wing-pads or short or long wings may be found. Nymphs usually have fewer antennal segments than adults.

Psocids or booklice

M. Herbut

Males may or may not be present and reproduction is by egg laying or by live birth. Each female lays 20-100 eggs, singly or in small groups. Nymphs molt five or six times before reaching the adult stage. Several generations occur each summer outdoors and reproduction can be continuous indoors.

Food Hosts and Damage

Psocids feed primarily on molds and mildew and are found in damp secluded places, both indoors and outdoors. Their foods are chiefly molds but they are known to also feed on cereals, starch, pollen, fragments of dead insects, paste in bookbindings, and similar animal and vegetable matter. They cannot bite man or animals and are harmless except for the contamination of food and any damage they may cause in feeding upon moldy paper such as books and wallpaper. At times they are a nuisance because of their abundance, spreading throughout buildings.

Psocids may reach very high numbers in damp grain, probably feeding on molds in the grain. They are not believed to cause damage.

Controls

Warm, dry and well aerated conditions can reduce psocid populations by preventing mold growth and hence removing the food source. A thorough house cleaning with sunning, drying or airing of infested rooms and materials provides good control. The use of a household aerosol insect spray or other chemical may be used to control psocid populations indoors if the source cannot be found. Turning and drying tough grain will not only discourage mold growth but also mechanically kill many psocids by crushing.

Thysanura
Firebrat
Thermobia domestica (Parkard)

Silverfish
Lepisma saccharina Linnaeus

Appearance and Life History

The firebrat and silverfish are introduced insects that have adapted well to heated buildings. Their bodies are slender, wingless, flattened and covered with scales. Firebrats are mottled silvery-grey in appearance; silverfish are more uniformly silver in color. The body is wider at the front and tapers to a somewhat pointed tail. Adults vary from 8 to 13 mm in length with the nymphs being similar in shape but smaller. On the head are two long antennae. Three tail-like appendages protrude from the hind end. These feelers or cerci are quite stiff and are sensory in function. The center feeler in firebrats is directed straight back from the tip of the body while the two lateral feelers are usually held at right angles to the middle feeler. The two lateral hind feelers of silverfish are directed backwards at a sharp angle.

Adult silverfish

H. Philip

They are both very active at night and hide during the day, therefore rarely seen. When objects under which they are hiding are moved, they dart about quickly, stopping at short intervals, and then moving on rapidly to seek other hiding places. Their three pairs of well developed legs allow them to run very fast. They cannot climb smooth surfaces, so are often trapped in bathtubs, wash basins or glass trays.

Firebrats live in hot dark places such as around furnaces, fireplaces and in insulation around hot water and heat pipes. They may follow furnace pipes to rooms where they may find food. Temperatures of 32-40°C and relative humidity of 70-80 percent are preferred for rapid growth and population build-up. Their common name is well deserved when one considers that they are a nuisance in extremely warm locations. Silverfish like lower temperatures (22-27°C) and somewhat higher humidity (75-95 percent).

Eggs are laid in batches of about 50 at a time in cracks and crevices near their food supply. Eggs are soft, white and opaque when laid and later turn yellowish. Eggs hatch in about 2 weeks. Nymphs look identical to adults in shape but may be lighter in color and smaller in size. Nymphs may reach sexual maturity in from 3 to 24 months depending on temperature and humidity. Molting does not cease when adulthood is reached but continues for another 2-3 years. An adult female may therefore double her size and go through 40-60 molts in her life time. She will molt and mate after each batch of eggs is laid. Temperatures below 0°C and above 44°C are usually fatal.

Food Hosts and Damage

Foods high in plant starches and sugars are favored. Cereals, moist wheat flour, paper on which there is glue or paste, sizing in paper (including wall paper), book bindings and starch in clothing are the most common food items. Bakeries, with their warm environment and ample starchy flour, make ideal homes for firebrats. Both are attracted to bonding glues used in wood processing plants and have then been carried to new building sites on wall boards. They have been reported to attack a number of synthetic fabrics, such as rayon, besides the natural fibers of cottons and linen. They rarely feed on wool, hair or other material of animal origin. Scales and excrement may leave yellowish stains on fabrics and paper. They may eat out and destroy book bindings and the glazing of high-quality paper, leaving irregular patches and notches on pages in books.

Control

Cultural control involves changing the environment to discourage development. This may include reducing temperatures or changing lighting patterns in infested areas. Improved sanitation and removal or proper storage of food sources is also helpful. Residual insecticidal sprays or dusts may be used under certain conditions depending on the surface treated, exposure of the surface to people and pets, length of control desired and type of application required (spot or crevice application).

Insect Pests of Farm-Stored Grain

Acari
Grain Mites
various species

Appearance and Life History

Grain mites are very small (0.3-0.6 mm long) wingless creatures with white to brown, soft, translucent bodies with four long hairs arising from the hind end, and the characteristic eight legs and two indistinct body regions. Eggs are round, clear and small. The larvae have only six legs. Two or three nymphal growth stages having eight legs occur, with one of these often being a non-feeding stage known as a hypopus.

Grain mite on stored canola seed

M. Herbut

Grain mites can multiply very rapidly as each female can lay between 500 and 2500 eggs in a month. Under favorable conditions of food, temperature (20°C) and grain moisture (14 percent), development from egg to adult takes only about 2 weeks. The non-feeding hypopus stage may survive for months as it is resistant to low temperatures, drying, starvation and most fumigants. Adults and immature stages, except the hypopus, die when exposed for a week to temperatures below -18°C. Mites will remain active at grain temperatures above 4°C and grain moisture levels above 13 percent.

On the prairies, about eight species of mites inhabit stored grain with the grain mite, *Acarus siro* Linnaeus, being the most damaging. The longhaired mite, *Lepidoglyphus destructor* (Schrank), is the most common species and may be recognized by its long stiff hairs and rapid, jerky walk. A beneficial species, the cannibal mite, *Cheyletus eruditus* (Schrank), feeds on grain mites, longhaired mites and insect eggs. Their numbers are usually not high enough to completely control mite infestations.

Food Hosts and Damage

Except for the predatory mites that feed on other mites, grain-inhabiting mites feed on the germ of grain kernels causing germination loss, and spread fungi (molds) which are also eaten. They prefer tough or damp grains and may feed on the whole seeds of cereals, oilseeds and weeds, or the cracked and broken pieces of grain along with flour and grain dust. Heavily infested grain becomes tainted with a disagreeable odor and unpalatable as animal feed, often causing digestive disorders in some animals. Their feeding and excrement can cause the growth of molds and the grain to heat, creating hot spots within the grain mass. These hot spots are susceptible to increased mite activity and more damage.

Control

Bins should be cleaned and sprayed with an insecticide as a preharvest safeguard. New grain should not be stored on top of old grain which might be infested. Turning the binned grain several times to mechanically crush the mites is an effective control method. Aerating grain to lower the grain temperature and reduce moisture content to below 12 percent will help to keep an infestation in check. Screening and fanning infested grain will remove and kill most of the mites including the eggs. Fumigation is not effective against mite eggs and hypopus stages.

Coleoptera
Confused Flour Beetle
Tribolium confusum Jacquelin du Val

Red Flour Beetle
Tribolium castaneum (Herbst)

Appearance and Life History

Confused and red flour beetles are very similar in life history, habits and appearance. Adults are small, reddish-brown beetles about 4 mm in length and somewhat flattened. Antennae of confused flour beetles gradually enlarge towards the tips, ending in clubs consisting of four enlarged terminal segments. Antennae of red flour beetles end abruptly in clubs consisting of three segments. Confused flour beetles do not fly while red flour beetles are capable of flight.

Adult confused (upper) and red (lower) flour beetles

H. Philip

The small white eggs are coated with a sticky secretion and become covered with flour or meal, and can readily adhere to the sides of sacks, boxes and other containers. Larvae are small, cylindrical and yellowish-brown with dark heads and a pair of slender, pointed processes at the hind end. Mature larvae may reach a length of 6-8 mm. The naked pupae gradually change from white to yellow to brown before transforming into adult beetles.

Female confused flour beetles lay 200-700 eggs over an 8-month period while female red flour beetles lay 300-400 eggs over a 5-month period. Eggs are laid loosely in or on food material. Larvae hatch in about a week and feed for several weeks to over 3 months depending upon the temperature, humidity and food supply. Pupation for both species lasts 8 days and takes place in the food material.

Development from egg to adult may be completed in 15-20 days under optimum conditions of 32°C and 70-90 percent relative humidity. Under grain bin conditions, the life cycle can take 6 weeks or longer in cold weather. Four to five generations may develop in a year, with large numbers of adults present because of overlapping generations. Adults can live up to 2 years. The temperature range for development is 20-40°C.

Food Hosts and Damage

Both species of flour beetles are common pests of stored plant products in houses, granaries, grocery stores and flour mills. Food products attacked include flour, cereal products, peas, beans, shelled nuts, dried fruits, spices and mild chocolate. Confused flour beetles feed on cracked, milled or damaged grain as they are unable to penetrate sound kernels. They are most commonly household pests but can live in flour mills and feed mills. The red flour beetle is an important pest of farm stored grain and if given the proper environment, can reproduce rapidly. It prefers damaged grain, but will attack whole wheat, feeding first on the germ and then on the endosperm. Adults of both species release a malodorous secretion that renders milled products unfit for consumption.

Another common species of flour beetle that also infests household stored products is the large flour beetle, *Tribolium destructor* Uyttenboogaart. This insect is dark brown to black in

color, 6-7 mm in length, and has a similar life cycle to that of the confused and red flour beetles.

Large flour beetle adult and larva

M. Herbut

Control

Prevention of infestations is the cheapest and least troublesome of all control practices. Sanitary practices include regular cleaning of cupboards, bins, milling equipment and any area where grain dust or crumbs may accumulate. Food hosts should be inspected for contamination before being stored. Tight fitting containers should be used in the house. Grain bin walls should be sprayed with an insecticide before grain is placed into storage. Infested food may be frozen in a freezer for 4 days or heated in an oven at 55°C for at least 30 minutes to kill all stages of the beetles.

Red flour beetles may be killed during movement of the grain by turning the grain and moving it from one bin to another. Cooling the grain to -7°C or lower for 6 weeks has been effective. Infestations can be controlled by treating the grain with a recommended fumigant or insecticide.

Granary Weevil
Sitophilus granarius (Linnaeus)

Rice Weevil
Sitophilus oryzae (Linnaeus)

Appearance and Life History

The granary and rice weevils are pests of whole stored grain and can easily be recognized by their downward curved, slender snout with chewing mouthparts at the end. Adults are cylindrical in

shape with shiny brown to black wing covers. Antennae are elbowed. Whitish eggs are small, soft and oval. Larvae are white legless grubs with brown heads. Pupae resemble adults but are white and soft-bodied.

The granary weevil is uniformly reddish to black in color with large oval pits on the thorax and wing covers, is 4-5 mm in length, and lacks functional wings, hence is unable to fly.

Adult granary weevils feeding on wheat kernel

M. Herbut

The rice weevil is reddish-brown in color with two light red or yellow spots on each wing cover, with small, round, closely compacted pits on the thorax and wing cover, 3-4 mm in length, and with functional wings.

A female weevil chews a hole through the seed coat and deposits a single egg in the cavity. She then seals the egg into the kernel with a gelatinous secretion. The egg hatches in 3-5 days and the larva begins feeding in the kernel. There is only one larva in each infested kernel. The larva completes its growth, pupates and develops into an adult weevil within the kernel. After reaching the adult stage, it eats its way out of the kernel, leaving a neat round exit hole.

Adult weevils may live 4-8 months and each female may lay 150-400 eggs during this time. The entire life cycle may be completed in 4 or 5 weeks in warm weather. Weevils can survive cool winter months, however the life cycle will take much longer to complete. By passing most of its life within the kernel, the insect is protected from many natural enemies as well as from sudden changes in temperature and moisture.

Food Hosts and Damage

Grain weevils feed on and in stored grains such as wheat, oats, barley, rice, corn and other grain products. Larvae completely destroy the kernels, leaving only the hulls. Infested grain may not have any visible signs of damage but infested kernels can be detected by staining techniques which dye the egg plugs. Adult feeding on the bran often prepares the way for other insect species that feed on damaged kernels but are unable to penetrate sound grain. Weevil feeding activity also contributes to grain heating and further damage. Field infestations of these weevils do not occur on the prairies but both are encountered in stored grain as primary invaders. Since the granary weevils are flightless, they are restricted to grain storage bins that can be reached by walking or to which they are transported by man. Rice weevils are occasionally found in store-bought processed rice.

Control

Preventive measures which keep the weevils out of stored grain are the most effective controls. These may include removal of waste grain and feed, cleaning grain storage structures and areas around them, cleaning grain-harvesting equipment and spraying grain bins with an approved insecticide before filling them. A heavy infestation can be treated with an approved fumigant or by cooling the grain to prevent weevil reproduction and damage.

Rusty Grain Beetle
Cryptolestes ferrugineus (Stephens)

Appearance and Life History

The rusty grain beetle is a flat, reddish-brown insect about 2 mm long. This species can be distinguished from flour beetles by their smaller size and long, slender antennae that are about one-half the length of the body. Eggs are white and elongated. Larvae are slender, up to 3 mm long when mature, white and have two brown projections at the tail end. Pupae are white, soft-bodied and resemble the adults with reduced wings.

Adult rusty grain beetles

M. Herbut

Rusty grain beetles are very active insects, running quickly through warm grain and flying when the air temperature is above 25°C. Females lay 200-500 eggs each either in cracks in grain kernels near the germ region, or between the kernels, or among grain dust and debris. The eggs hatch in 4-5 days, depending on the temperature, and larvae commence feeding on the germ and endosperm. After feeding, the larvae pupate inside the kernel, emerging in 3-9 days. Mating occurs 1 or 2 days after adult emergence and egg laying begins shortly thereafter. Egg laying may last for 1-9 months with females laying up to 7 eggs per day.

The life cycle from egg to adult may take only 4 weeks under conditions of 15 percent grain moisture content and temperatures of 32°C. Beetles do not develop in dry grain with a moisture content of less than 12 percent or when the relative humidity is less than 40 percent. Temperature range for complete development is from 18 to 40°C. Development under less than ideal conditions may take over 3 months.

Food Hosts and Damage

The rusty grain beetle is one of the most serious pests of stored grain on the prairies. They prefer wheat and rye over other grains. Adults feed on damaged kernels, grain dust and on sound kernels that have been exposed to a long period of high humidity and temperature. Overheated, high moisture grain provides the optimum environment for this stored grain pest. High populations can affect seed germination due to the feeding habits of the larvae on the germ. Heavy infestations also cause the grain to heat and spoil and the spread of fungal spores in the stored grain.

Since rusty grain beetles are able flyers, they disperse readily from infested to uninfested granaries. Adults have also been found under the bark of trees where they overwinter if no stored grain is available. Due to their small size and flattened body, they find little problem in entering granaries filled with grain.

Rusty grain beetles can be most serious in years of continuous wet harvest periods, very warm temperatures at harvest times for several years in a row, and periods of overproduction or lack of sales with grain being stored for long periods, especially in temporary storage facilities.

Control

Sanitation procedures to prevent infestations include thoroughly cleaning empty grain bins and spraying them with an insecticide before loading new grain into them. New grain should not be piled on old grain. The use of a grain protectant mixed with grain going into storage will help prevent an infestation from developing. Drying and cooling grain shortly after harvesting are valuable precautions. Infestations in bins can be controlled by treating the grain with a recommended fumigant or, in winter, by cooling the grain to -7°C and holding it at that temperature for at least 6 weeks. Low temperatures may not kill the rusty grain beetle but development will cease and damage to grain will stop.

Yellow Mealworm
Tenebrio molitor Linnaeus

Appearance and Life History

The yellow mealworm is the largest and most easily recognized insect that infests stored cereal products. Adult beetles are dark brown or shiny black, about 15 mm long, with finely grooved wing covers. The white bean-shaped eggs are coated with a sticky secretion to which flour, meal or grain dust adheres. Young larvae are white but gradually turn yellow as they mature. Mature larvae measure up to 30 mm long, are slender, smooth, shiny-bodied, yellow to brown in color and closely resemble wireworms. The areas between the segments are lighter in color, thus giving the darker segments the appearance of rings around the body. Pupae are 13-20 mm long, white when first formed but gradually turn yellowish-brown in color.

Yellow mealworm adult and larva

M. Herbut

Mealworms overwinter in the grain as larvae which pupate in late spring or early summer. Adults emerge in about 2 weeks, and after mating, females deposit their eggs, either singly or in small clusters, in the food material or on the sides of the bins. Each female lays 300-500 eggs over her life span of 2-3 months. Eggs hatch in 4-7 days at normal spring or summer temperatures.

The larvae may become full grown in about 3 months, however, they continue to feed and molt, but growing very little, for at least a further 9 months. Larvae are resistant to dryness and cold, and can survive several weeks at temperatures as low as -15°C. There is usually one generation per year but 2 years may be required under unfavorable conditions.

Food Hosts and Damage

Yellow mealworms feed in and around grain bins, particularly in dark, moist areas where the grain has not been disturbed for some time. Both adults and larvae prefer damp, decaying grain or milled cereal products. They are found wherever accumulations of grain occur, such as in neglected corners of feed mills, under bags of feed in warehouses and bins containing damp grain. Lone adults, which may be found at any time of the year, do little or no damage but full-grown larvae can do serious injury to whole grains when held for long periods without being moved. They also feed on meal, meat scraps, feathers and dead insects.

Control

Proper sanitation is essential for the effective control of yellow mealworms. Thorough and regular cleaning of all storage areas to remove old grain, meal, flour, sweepings and refuse followed by a residual spray to control any insects remaining after the cleaning will prevent infestations. Areas under loading docks or warehouse floors through which grain dust and debris settles must also be cleaned. Fumigation is rarely required for this insect.

Insect Pests of Humans and Domestic Animals

Acari
Parasitic Mites
various species

Appearance and Life History

Mammals and birds are attacked by several economically important species of parasitic mites. Some species spend their entire life cycle on the host whereas others only visit their hosts to feed. Adults are generally very difficult to see without the aid of a magnifying lens because they are only 0.4-1.0 mm long. They have four pairs of legs, two body sections, and vary in color from white to dark red to black, depending on the species. Eggs are oval and white but are rarely seen. Larvae have only three pairs of legs; nymphs have four pairs and resemble the adults in appearance except are smaller in size.

The life cycle of parasitic mites includes four stages: egg, larva, nymph and adult. The duration of the individual stages, time to complete a life cycle and the behavior of the mobile stages vary according to the species.

The sarcoptic mange or itch mite, *Sarcoptes scabiei* (DeGeer), is an extremely small (0.4-0.6 mm) mite that burrows into the outer layers of skin of mammalian hosts. Females deposit their oval eggs at 2- to 3-day intervals over a 2-month period as they burrow through the skin. Larvae hatch in 3-5 days and either burrow in the skin or more commonly move freely over the skin or hide in hair follicles. Larvae molt into nymphs within 2 days, and the nymphs, also free-living on the skin, soon transform into either males or immature females. Male mites remain on the skin where they mate with the immature females. Fertilized females then burrow into the skin to start feeding and laying eggs. The entire life cycle from egg to mature female takes 10-14 days.

The chorioptic mange mite, *Chorioptes bovis* (Gerlach), resembles the sarcoptic mite but has longer legs. The entire life cycle is spent on the skin. The time to complete the life cycle is about the same as for sarcoptic mites.

The psoroptic mange mite, or sheep scab mite, *Psoroptes ovis* (Hering), is not very common on the prairies. The life history is the same as for the chorioptic mite. The active stages do not burrow but may be concealed by scab formations resulting from their feeding activity.

Demodectic mange or follicle mites in the genus *Demodex* are very small (0.25 mm long), slender, elongate mites that pass their entire life cycles on the skin of the host. Their shape allows them to live in hair follicles and skin glands. The life cycles require 18-24 days to complete.

The scalyleg mite, *Knemidokoptes mutans* (Robin and Lanquetin), is similar in habit and life history to the sarcoptic mite.

The chicken mite, *Dermanyssus gallinae* DeGeer, measures 0.4 mm long, is greyish when unfed but red to almost black when fed. These mites only visit the host to feed at night; the remainder of the time is spent hidden in cracks and crevices in the vicinity of the host. After feeding, fertilized females deposit batches of 25-50 eggs in cracks and crevices. Eggs hatch in 1-2 days and the subsequent larvae and nymphs feed for about a week before maturing into adults. The entire life cycle can be completed in 10 days under favorable conditions. Chicken mites can survive starvation for 4-5 months during the summer and even longer in winter.

The northern fowl mite, *Ornithonyssus sylviarum* (Canestrini and Fanzago), is slightly larger than the chicken mite as adult, is oval to elongate in shape and dark red to black in color. The entire life cycle, which takes 5-7 days to complete, is spent on the host. After feeding, fertilized females lay 2-5 eggs at a time in the fluff of the feathers. Eggs hatch in 1-4 days into a non-feeding larval stage which soon molts into the first nymphal stage which feeds. Within 2 days, the second nymphal stage emerges, does not feed, and in less than a day transforms in the adult stage. This species can live off the host for 2-3 weeks under ideal conditions. During heavy infestations, the mites can be found in the nest, on roosts, in cracks and on eggs and walls.

Food Hosts and Damage

Sarcoptic mange mites or itch mites attack humans, livestock and dogs. Populations existing on one host species are not readily transferred to another, and a varietal name is often added to *Sarcoptes scabiei* to indicate the host animal (for example, *S. scabiei* variety *bovis* for cattle mange mite; *S. scabiei* variety *sius* for hog mange mite). The burrowing activity of the females as they dissolve skin tissue and ingest the fluids causes intense itching and irritation. Small papules or red spots appear where the mites are burrowing and are capped off by small yellow granules of dried serum. In later stages, the skin appears dry, scurfy, rough, red and thickened. Constant rubbing and scratching will result in loss of hair, scab formation and possibly secondary infections. In cattle, mange infestations are first noticed inside the thighs, the brisket and around the base of the tail. In pigs, infestations start in the ears, around the nose and eyes, and on the neck. The neck, head and shoulders are the first areas infested in horses. If left uncontrolled, the infestations can spread over the rest of the animal's body. Uncontrolled infestations can result in reduced animal productivity, emaciation, and, in some cases, death. Infestations spread by direct animal to animal contact and by animals coming in contact with surfaces recently rubbed upon by mangy animals. Infestations in livestock are most severe during the winter months.

Hereford bull infested with sarcoptic mange mites

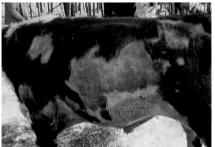

Alberta Government Photo

Sarcoptic mange infestations of humans is called scabies or "the itch" and signs of infestations are as previously described. Infestations generally start between the fingers, in the bend of knee and elbow, on the penis, under the breasts or on the shoulder blades. An infestation may progress for a month before the person experiences any itching. Mange mites are transferred through intimate personal contact. Common mange of dogs appears first on the muzzle, around the eyes, on ears and breast, and gradually extends over the rest of the body. Due to constant scratching and nipping, hair is lost, the dog may become emaciated or secondarily infected, and even die of exhaustion from the itching and reaction to the infestation.

Chorioptic mange mites feed on cattle, horses, sheep, goats and rabbits. Because these mites do not burrow, infestations are much less debilitating than those of sarcoptic mange mites. However, the feeding activity of the mites as they puncture skin tissue does irritate the animals, causing rubbing and scratching. Scabs will form around which the mites will feed. Infestations usually remain localized on the lower legs, tail or neck. Mites can be transferred as previously described for sarcoptic mange mites.

Psoroptic mange mites feed on cattle, horses, sheep and goats, and are the least common of the three mange mite species on the prairies, however, they have been a severe problem in the U.S. on cattle. These mites feed at the base of hairs, piercing the skin and causing inflammation and scab formation. Hair loss due to rubbing and scratching and reduced animal productivity can result from uncontrolled infestations. In sheep, the condition can result in reduced wool quality and quantity. Mites are transferred as previously described for sarcoptic mange mites.

Demodectic mange mites are commonly called follicle mites because they live in hair follicles and associated glands. Separate species of *Demodex* attack cattle (*D. bovis* Stiles), horses (*D. equi* Railliet), goats (*D. caprae* Railliet), hogs (*D. phylloides* Csokor), cats (*D. cati* Mégnin), dogs (*D. canis* Leydig) and humans (*D. folliculorum* (Simon)). Demodectic mites rarely adversely affect the health of the hosts. In extreme cases, dermatitis, localized swellings or papules and patches of hair loss may occur in livestock. Most humans are believed to harbor follicle mites, especially around the nose and eyelids. Red mange of dogs is caused by *D. canis* in association with a staphylococcus bacterium.

Scalyleg mites burrow in the skin of the legs of chickens, turkeys, pheasants and other birds. Infestations cause the condition known as "scaly-leg", recognized by the peculiar scabby enlargements and encrustations on the legs and feet. Birds can become crippled and thus less productive. This condition is restricted to birds allowed access to the ground or roosts, and thus is not a problem in caged-layer facilities.

Chicken mites (roost mites) can be a very serious problem in poultry operations. Because these mites feed on blood, heavy infestations can cause birds to become droopy and weak and have pale combs and wattles, resulting in reduced egg production and poor feed utilization. Young chickens and setting hens may die from blood loss. Infestations can be detected by examining the birds at night when the mites are feeding or during the daytime by inspecting cracks and crevices for masses of eggs, mites, black and white mite excrement and silvery cast skins. Infestations are generally most severe in the summer.

Northern fowl mites live primarily on laying chickens, congregating near the vent, tail, back and neck, sucking blood and irritiating the birds. High infestations can reduce egg production and food intake, and cause weight losses, anemia and death. Egg handlers have also been bitten by these mites. Heavily infested roosters have a reduced ability to mate. Infestations are most severe during the winter months.

Control

Various chemicals are registered for the control of parasitic mites of humans and domestic animals. Infestations of sarcoptic and psoroptic mange of cattle must be reported to veterinarians as they are quarantinable diseases under federal legislation. Any confirmed mangy cattle and any cattle in contact with them, must be quarantined until declared free of mange following prescribed chemical treatments. Even if chorioptic mange is suspected, it is advisable to have a veterinarian make a diagnosis to ensure that sarcoptic or psoroptic mange mites are not present. For other livestock and poultry parasitic mite control information, consult your local veterinarian or agricultural extension office for the latest recommendations. Human scabies is controlled by application of acaricidal lotions and creams.

Cultural control of parasitic mites includes isolation of affected animals until disinfested, and thorough cleaning and disinfection of pens, roosts, alleys, fences, floors, walls and other surfaces to remove mites and mite eggs before introducing new animals, especially if mange has been present.

Ticks
various species

Appearance and Life History

Ticks are eight-legged external parasites of mammals and birds. Unlike insects, which have three body regions, ticks have only two regions – the mouthparts and the general body region that bears the legs and unsegmented abdomen. Adult ticks are flattened, ovoid and measure up to 5 mm in length. Males have a hard plate covering their entire back, whereas in females the hard plate covers only a small portion of the front end. Eggs are ovoid, variously colored and covered with a sticky substance that holds them together and prevents drying. Larvae or seed ticks have only six legs; nymphs have eight legs and resemble adult ticks but are smaller. The life cycle of ticks includes four stages: egg, larva, nymph and adult. Time required to complete each stage and total life cycle differs according to species, food supply and environmental conditions.

The Rocky Mountain wood tick, *Dermacentor andersoni* Stiles, also known as the wood tick or paralysis tick, is found in open, low bush or scrub-covered grasslands of south-central and southern Alberta and eastwards to mid-Saskatchewan. The body, head and legs of the female tick are dark brownish-red in color, the shield being white with small curved red markings on each side of the centre. The male is greyish-white with a number of irregular dark, blue-grey markings. After mating and feeding in the early spring, each female will lay up to 6500 brown eggs in a single mass under ground debris, then die. Eggs begin hatching in 4-6 weeks and may continue for 2 weeks or more. The larvae or seed ticks usually begin to appear in June and are most numerous in July when they climb grass to await suitable hosts upon which to feed for about 6 days. Engorged larvae drop from the host and molt into nymphs 1-4 weeks later. If temperature and

host-availability are favorable, nymphs will feed, molt and overwinter as adults. Under hot conditions, nymphs will not seek hosts and will overwinter unfed. With the onset of warm spring weather, overwintered nymphs become active, climbing grass from which to attach to passing small mammals. They will feed for about a week, and then drop to the ground where they remain inactive for 6-10 weeks. During this inactive period, nymphs molt into adults which do not feed until the following spring. Depending on weather conditions, overwintered adult ticks can become active in March and remain active through June. They climb tall grass and wait, head downwards, for any suitable host animal to brush past them. Mating occurs while the females are attached and feeding. Engorgement usually takes a week to accomplish. Male ticks do not enlarge after feeding, however, female ticks will swell to 12 mm or more and assume a bluish-grey color. If no suitable host is available, adult ticks will seek sheltered sites once again to avoid the high summer temperatures and to overwinter. The life cycle requires three hosts and 1-3 years to complete, depending on host availability.

Female (left) and male (right) Rocky Mountain wood ticks

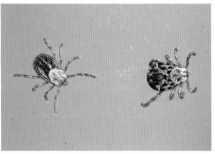

M. Herbut

The winter tick, *D. albipictus* (Packard), also known as the horse tick or moose tick, is slightly larger and paler in color than the Rocky Mountain wood tick. The white shield of the female has a long central red line in addition to the two lateral ones. The shield covering the back of the male is whitish with numerous irregular salmon-colored markings. Eggs are laid on the ground in a large mass over a period of several weeks in early summer, about 200 being laid each day until a total of 4000 or more

have been deposited. Hatching occurs in 3-6 weeks, and the young larvae remain inactive under soil litter until the fall when they climb grass and low shrubs and wait to attach to a passing host. Several hundred larvae will congregate on a single grass blade or bush twig. Larvae will remain on grass and shrubs through snow falls, attaching to hosts on sunny days. Once attached to a host, larval and nymphal feeding and molting take place without leaving the host. Adult ticks are evident on hosts about 6 weeks after larvae attach. Females feed continuously for a week, during which time mating occurs. Although engorged mated females drop off the host, egg laying is delayed until spring. Late-feeding females may still be attached when Rocky Mountain wood ticks become active in early spring. Winter ticks are active from about October to April, and the life cycle is completed in one year on a single host. This tick is mainly confined to the wooded and forested regions of the prairies.

The American dog tick, *D. variablis* (Say), is less common than *D. andersoni* and *D. albipictus* in Alberta, but can be abundant in Manitoba and parts of Saskatchewan. It closely resembles *D. andersoni* in appearance, habits and life history, but can be distinguished by the larger red marking down the centre of the white shield of the female. Adults are active in the spring at which time egg laying takes place. Larvae and nymphs are active during the late spring and summer; both nymphs and adults overwinter. The life cycle requires three hosts and 2 or 3 years to complete.

Food Hosts and Damage

The mouthparts of ticks are modified for cutting host skin and sucking up blood. The structure of the mouthparts also aids in securing the ticks to the host, making detachment difficult during their extended feeding periods. While feeding, saliva containing anticoagulants and other substances is injected into the wound to allow continuous ingestion of blood.

Rocky Mountain wood tick larvae and nymphs feed on marmots, moles, mice, squirrels, ground squirrels, rabbits and other small mammals. Adults feed on large mammals such as cattle, sheep, horses, dogs, deer, elk and humans. "Tick paralysis" can occur in humans as a result of female tick feeding. It is characterized by flaccid ascending paralysis that can cause death by respiratory failure. Recovery is rapid if all ticks are removed before the paralysis reaches its final stages. Tick paralysis in livestock is very rare on the prairies but can be a major problem to cattle producers in southern British Columbia. These ticks are also vectors of tularemia, a widely distributed plaque-like disease of rodents, occasionally contracted by man and some domestic animals. Another important, usually fatal disease transmitted by this tick to humans is Rocky Mountain spotted fever, however, cases have been rare on the prairies in recent years.

Moose, deer and horses are the main hosts of the winter tick, with cattle and elk occasionally attacked. Severely infested animals can suffer anemia because of the amount of blood lost to tick feeding. The combination of tick anemia and severe winter weather and poor food supply can be fatal. Horses on wooded range pasture during the fall and winter are most susceptible to attack.

Dogs and coyotes are the preferred hosts of the American dog tick, however, cattle, horses, raccoons and humans can also be attacked by adult ticks. Larvae and nymphs feed on small mammals similar to the Rocky Mountain wood tick. The dog tick is of minor economic importance to livestock and dogs, and is only an occasional annoyance to humans.

Human reactions to tick bites include inflammation, itching, swelling and ulcerations at the site of the bite. Improper or partial removal of the tick mouthparts can lead to further skin ulceration and secondary bacterial infections.

Control

Because winter ticks, Rocky Mountain wood ticks and American dog ticks are not sufficiently abundant on prairie livestock to cause economic losses except in isolated cases, producers are not required to maintain constant vigilance for their presence. Chemicals are registered for application to livestock to prevent or control infestations. Horses wintering on wooded range should be checked from December through February for the presence of winter ticks, especially if the pest has been previously present and food supply is poor.

People who visit areas where Rocky Mountain wood ticks or American dog ticks are present, especially during the period of adult tick activity, should examine themselves for ticks immediately upon leaving the area. Feeding ticks can be removed safely by holding a hot object such as a lit match near the tick. The tick will withdraw its mouthparts and be easily removed without risk of leaving the mouthparts embedded in the skin. Pets should also be examined for ticks, both nymphal and adult, if brought into tick-infested areas. Repellants can be used as protection against ticks.

Anoplura, Mallophaga
Lice
various species

Appearance and Life History

Lice are divided into two groups, the sucking lice and biting lice. Both groups are external parasites of birds and mammals. Adult sucking lice are soft-bodied, somewhat flattened wingless insects, varying in color from white to dark blue-grey and in length from 1.5 to 4.8 mm. Most are elongate in shape, the head usually pointed and narrower than the rest of the body. Nymphs resemble adults except that they are smaller in size. Eggs, or nits, are cream-colored, ovoid and about 1-1.5 mm long. Adult biting lice are very small, wingless, somewhat flattened insects, varying in color from white to reddish-brown and in length from 1.2 to

3.0 mm. The head is large in proportion to the body and is broadly rounded in front. Eggs are similar in shape, size and color to those of sucking lice. Nymphs resemble adults but they are smaller in size and softer bodied.

The life cycles are similar for both groups of lice. Females cement eggs singly to hairs or feathers close to the skin of the host where the warmth from the host aids in egg development. The eggs hatch in 10-20 days and the newly hatched nymphs immediately start feeding on the host. Lice go through three nymphal instars in 2-3 weeks before reaching the adult stage. New females will commence egg laying within 2 weeks and deposit a few eggs each day of their 4- to 5-week life span. The life cycle of biting lice requires 15-22 days to complete as compared to 18-34 days for sucking lice. Some species of biting lice have no males and reproduce parthenogenetically. Lice spend their entire life cycle on their hosts, and can survive for only a few days away from the host.

Food Hosts and Damage

Sucking lice feed exclusively on mammals, using their piercing and sucking mouthparts to obtain blood meals. Biting lice attack both mammals and birds, using their chewing mouthparts to feed on dry skin scales, dead scab tissue, feather parts and other skin debris. Most lice are host specific and limited to a single host species or closely related hosts. Sucking lice are less active on the host than biting lice which readily move about in search of food. Sucking lice are more debilitating because of their blood-feeding behavior. Several species of sucking and biting lice attack humans and domesticated animals.

The cattle biting louse, *Bovicola bovis* (Linnaeus), is a small louse with a reddish-brown head and light cream or yellowish-white body. These lice attack cattle, and preferred sites are the tail base, shoulders and top line of the neck and back. Because of their mobility, infestations can spread over the entire body, and their presence causes cattle to rub and scratch resulting in open sores, hair loss and rough haircoats. This species can reproduce parthenogenetically. Infestations are heaviest during the winter. The horse biting louse, *B. equi* (Denny), resembles *B. bovis* in appearance and damage inflicted; preferred sites on horses are the head, mane, tail base and shoulder area.

The goat biting louse, *Bovicola caprae* (Gurlt), attacks goats and resembles *B. bovis* in appearance and damage. The sheep biting louse, *B. ovis* (Schrank), attacks domestic and bighorn sheep. Wool of infested sheep becomes ragged and will be downgraded. Infestations are highest in the winter months.

The chicken body louse, *Menacanthus stramineus* (Nitzsch), attacks domestic chickens and turkeys, guinea fowl, pea fowl, quail, pheasants, ducks and geese. Heavy infestations of these pale yellow or brown lice may result in significant irritation and decreased performance. They puncture soft quills near their base and feed on the birds' blood. Other species of biting lice which may be found on chickens are the wing louse, *Lipeurus caponis* (Linnaeus), and the shaft louse, *Menopon gallinae* (Linnaeus).

The shortnosed cattle louse, *Haematopinus eurysternus* (Nitzsch), a sucking louse about 3 mm long, is commonly found on mature cattle. Infestations are heaviest during the winter, and preferred sites are the top and sides of the neck, dewlap, back, base of horns and the tail base. Infested animals appear greasy, dirty with rough haircoats, and display sores, cuts and hair loss from constant rubbing and scratching to relieve the intense irritation. Self-grooming and a healthy immune response system aid in regulating lice populations. Cattle unable to self-groom or having a weakened immune response system due to disease, malnutrition or some other stress factor will become more heavily infested and suffer greater losses than infested healthy cattle. Uncontrolled infestations can cause anemia, reduced weight gains and lower milk production, abortions, reduced calf birthweights and decreased sperm counts in bulls with

scrotal infestations. Heavily infested cattle can succumb to secondary diseases that they would normally resist. Some cattle, called "carriers", have insufficient natural resistance to infestations and carry high numbers of lice year round which can be a source of reinfestation of the herd.

The longnosed cattle louse, *Linognathus vituli* (Linnaeus), is a sucking louse found most commonly on dairy cattle and young cattle. They are shorter and narrower than shortnosed cattle lice. Preferred feeding sites include the dewlap, shoulders, sides of neck, the cheeks and the rump, usually in dense, small colonies. Heavily infested young cattle show similar symptoms of infestation as that of the shortnosed cattle lice.

Adult cattle biting louse (left), longnosed sucking louse (center), and shortnosed sucking louse

H. Philip

The hog louse, *Haematopinus suis* (Linnaeus), is a sucking louse that attacks swine. Preferred sites are in the folds of skin behind the ears and between the legs. These lice are the largest of the sucking lice, measuring up to 4.8 mm long. Infested animals rub and scratch to relieve the irritation, causing the skin to become thick, cracked, tender and sore. Animals become restless, less profitable and more susceptible to diseases which may cause death.

Other species of sucking lice that may occur on domestic livestock include the horse sucking louse, *Haematopinus asini* (Linnaeus), the sheep sucking louse, *Linognathus ovillus* (Neumann), the goat sucking louse, *L. stenopsis* (Burmeister), and the capillate louse or little blue cattle louse, *Solenopotes*

capillatus Enderlein. Dogs are attacked by the dog biting louse, *Trichodectes canis* DeGeer, and a sucking louse, *Linognathus setosus* (Olfers). Cats may become heavily infested with the cat biting louse, *Felicola subrostratus* (Burmeister).

Humans are also hosts for three species of sucking lice: the crab louse, *Phthirus pubis* (Linnaeus); the body louse, *Pediculus humanus humanus* Linnaeus; the head louse, *P. humanus capitis* DeGeer. Crab lice are small (1.5-2 mm long), whitish, crab-like in appearance, and commonly found in the hair of the pubic region. These lice do not move about very much, but remain secured to pubic hairs using modified claws. Crab lice are most commonly transferred during intimate body contact, but also on clothing in locker rooms, towels, bedding and toilets. These lice cannot survive off the host for more than 1 day. Head lice

Adult crab or pubic louse

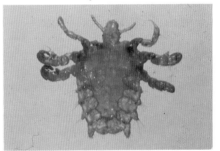

M. Herbut

and body lice are very similar in size (3-3.5 mm long) shape, color and life cycle. The major difference is in their habits. Head lice principally infest the head, but are occasionally found on hair on other parts of the body. Body lice occur chiefly on the clothing and move to adjacent body areas to feed. Head and crab lice cement their eggs on hair close to the skin whereas body lice lay their eggs in the seams of underwear and in places where the clothing comes in contact with the body. Head and body lice are readily transmitted under crowded and unsanitary conditions. Head lice infestations among school children has become a serious problem in many cities and towns. The bites of lice appear as reddened swellings that are itchy, and if the infestation is severe, the affected areas can become very

Adult human body louse

M. Herbut

sensitive and secondary skin infections can result from scratching to relieve the itchiness. The movement of head lice through the scalp also causes annoyance.

Control

For domestic animals, several chemical formulations and application methods are registered for the control of lice, and information on husbandry practices to prevent and control lice infestations is available from local agricultural extension offices or veterinary clinics. Lice are easily transferred among confined animals so it is important to be alert for lousy animals, especially "carriers", and to apply control procedures to prevent the spread and potential production losses. Products in the form of insecticidal shampoos and lotions are available to control human louse infestations. Avoiding contact with suspected lousy persons, clothing, bedding, or occupancy in sleeping quarters frequented by suspected lousy persons will eliminate the risk of becoming infested. School children who constantly scratch their scalp or complain of itchy scalp should be closely inspected for head lice. Fine-toothed metal combs, called "nit combs", are useful in destroying head lice nits or eggs.

Coleoptera
Lesser mealworm
Alphitobius diaperinus (Panzer)

Appearance and Life History

The lesser mealworm is frequently abundant in poultry manure and litter. Adult mealworms are dark brown to black, shiny smooth beetles about 6 mm long. Eggs are small, oval and white. Larvae are yellowish-brown, up to 15 mm long, with a distinct head and six small legs. Pupae are naked (not encased in a cocoon) and white in color.

Adult lesser mealworm

M. Herbut

Female beetles lay batches of up to 12 eggs in the litter and in cracks and crevices. Eggs hatch in 3-9 days; no hatching occurs below 21°C. Larvae require up to 4 months or more to complete their development depending on the temperature. Mature larvae pupate in the litter, cracks or crevices, or in building insulation. The pupal stage lasts 7-11 days. Adults can live 3-12 months. Winters are passed in or under the floors of poultry houses.

Food Hosts and Damage

Lesser mealworms can become extremely abundant in poultry litter where both adults and larvae feed on undigested and spilled grain, feathers and even dead or dying poultry. Foam insulation can be damaged by mature larvae tunneling through the material and chewing out large cavities in which to pupate. The beetles are also known to harbor various avian pathogens such as *Salmonella* and *Escherichia* bacteria and *Aspergillus* fungi. Adults fly only at night and are capable of dispersing from infested buildings or litter piles to nearby residences or other poultry barns.

Control

Both chemical and cultural means of controlling lesser mealworm are available, however neither approach will prevent reinfestations. Insecticides applied to the litter, floors and walls will provide short-term reduction in mealworm numbers. Beetle population build-up can be delayed by thoroughly cleaning the barns between batches of birds (including removal of litter and grain beneath floors and grain bins), and then applying a residual insecticidal spray to the floors.

Diptera
Black Flies
Simulium spp.

Appearance and Life History

There are several species of black flies on the prairies that attack livestock, poultry and humans. Adult black flies, also known as buffalo gnats or sand flies, are small (2-5 mm long), dark grey to black insects with small heads and a characteristic humpbacked appearance. Eggs are very small and yellowish in color. The larvae are slender, measuring up to 8 mm when mature, and are dark in color. They live on the bottom of streams and rivers attached by means of posterior suckers with small hooks to stones, submergent branches and vegetation, and to floating debris. They move in a looping or "inchworm" fashion or at the end of a silken thread which they extrude. They feed by means of two fan-like brushes which are extended from the sides of the head to filter suitable food materials from the passing water. Colonies of larvae appear as dense, slimy, moss-like coatings on the substrate. Pupae are small (3-5 mm long) and encased in slipper- or boat-shaped cocoons attached to submerged substrate.

Adult northern black fly

M. Herbut

Most economically important species of black fly adults emerge from submerged pupae during the first 2 weeks of June, however, some species do appear in May. In general, females mate and seek blood from hosts before returning to streams and rivers to lay eggs, either dropped singly on the water, or laid in compact masses on stones and vegetation. Each female may lay up to 500 eggs, most of which hatch in 6-12 days. The duration of the larval stage varies with species from a few weeks to almost 1 year. The pupal stage can take from a few days to several weeks before adults emerge, float to the surface of the water, and take flight immediately. Generally more than one generation is produced each year, and because two or more species may be found in one locality, black flies can be present from May to October.

Food Hosts and Damage

Only female black flies attack livestock and humans because of their need of blood proteins for egg development. The two most important species attacking cattle in the central and northern regions of the prairies are *Simulium arcticum* Malloch (the northern black fly) and *Simulium luggeri* Nicholson and Mickel. These species breed in large rivers and have caused significant cattle production losses, even deaths, during outbreak years. *Simulium arcticum* prefers large animals such as cattle, attacking around the head and lower body areas where the hair is less dense. Humans are rarely attacked. *Simulium luggeri* attacks primarily around the head of cattle and will attack humans. *Simulium vittatum* Zetterstedt (the striped black fly) females often feed in the ears of horses and cattle causing pain, sores and irritability. *Simulium venustum* Say (the whitestockinged black fly) readily attacks humans as well as livestock, and along with *S. luggeri*, are serious problems in recreational areas near black fly breeding sites. *Simulium rugglesi* Nicholson and Mickel and *S. venustum* are reported to transmit a blood parasite (*Leucocytozoon simondi*) to turkeys, ducklings, goslings and wild ducks.

For species attacking cattle, direct feeding damage appears as bleeding or scabbed-over sores around the eyes, on the brisket and underside, on the

scrotum of bulls and on teats and udders of cows. Cattle under attack from these voracious blood feeders refuse to graze leading to reduced weight gains and milk production. Cattle either bunch together for protection or seek shade under trees or in buildings to escape the flies. Breeding cycles can be interrupted when bulls ignore receptive cows in favor of escaping the flies. Cows with damaged teats refuse to nurse their calves due to the irritation, leading to reduced calf weaning weights. Calves born during high black fly abundance, or cattle not previously exposed to black flies and brought into an outbreak area, can die within hours or days due to the intense biting of the flies if they are not protected long enough to get used to the fly bites and saliva secretions.

Steer showing northern black fly feeding damage to brisket and around eyes and nose

M. Herbut

Weather conditions are important in regulating the abundance and activity of black flies. They are capable of traveling up to 20 km from their breeding sites in search of hosts. Winds can carry swarms up to 60 km or more from the breeding sites.

Control

Because of the swarming and feeding behavior of most economic species of black flies, protection of cattle, horses, other domestic animals, and humans is difficult. Access to shaded areas such as trees or buildings will provide areas for livestock to escape the flies. Burning old straw, hay or cleared brush piles to produce a smoke smudge provides relief for livestock. Insecticides applied as whole-body sprays or through backrubbers are effective in reducing

black fly biting and feeding activity, thereby allowing cattle to graze and bulls to breed uninterrupted. The calving period should be scheduled to end before May so that calves have a chance to get used to black fly bites, and to ensure that calves are not born during periods of high black fly activity. Cattle not previously exposed to black flies should not be introduced into infested pastures without first allowing the animals to develop tolerance to the presence of black flies. Repellants are available for human protection.

Bot Flies
various species

Appearance and Life History

Four species of bot flies are of economic importance on the prairies. Adult flies are robust, medium-sized flies (10-12 mm long) that somewhat resemble honey bees. They have nonfunctional reduced mouthparts. Eggs are about 1-1.5 mm long, cylindrical, and vary in color from yellow to black depending on species. Mature larvae measure 16-20 mm in length and resemble cattle grub larvae in appearance and color. Puparia are darkish and somewhat shorter than mature larvae.

Gasterophilus intestinalis (DeGeer) is the common horse bot fly. Each female lays 100-500 yellowish eggs while in flight, hovering around the host and darting in to deposit an egg. Eggs are firmly attached to hair on the inside of the forelegs, and occasionally on the outside of the forelegs, the flanks and the mane. The eggs incubate for 5 days before they can be stimulated to hatch by the warmth and moisture from the host's tongue. Under cool conditions, hatching can be delayed such that viable eggs can remain unhatched on hosts until autumn. The larvae spend a short time in the mouth before migrating to the esophageal area of the stomach where they attach to the walls and feed for 7-10 months. When mature, they pass out with the faeces and pupate in the top soil layer under the dung pile. Adult flies emerge in about 5 weeks, and are present from late July to late September.

Horse bot larvae attached to inner wall of the stomach

Gordon Surgeoner
University of Guelph

The female nose bot fly, *G. haemorrhoidalis* (Linnaeus) or red-tailed bot fly (because of its orange-red terminal segment), is a rapid flier and darts in to lay up to 160 blackish eggs singly on hairs on the upper and lower lips of hosts. Females are active June through August, and only live for about a week. Eggs hatch in a few days after some stimulation by moisture and friction from the host's tongue, and the young larvae burrow into the tongue or lips. They remain there for 6 weeks, then move to the pyloric region of the stomach or the duodenum and attach to the walls. In the late winter or early spring, mature larvae move to the rectum near the anus where they reattach for 2-3 days before dropping out to pupate under dung piles. Adults emerge in 3-8 weeks and the cycle is repeated. Egg-laying females are mainly active on bright sunny days.

Females of the throat bot fly, *G. nasalis* (Linnaeus), hover around the front of their host, darting in occasionally to lay up to 500 pale yellow eggs singly on the long hairs beneath the jaw of the host. Within 1 week, the eggs hatch without any stimulation, and the young larvae crawl along the skin to the host's mouth where they remain in the soft tissue for about 20 days. They then move down the digestive tract and attach in the same area as the nose bot larvae. In the spring the mature larvae pass out with the faeces and pupate under dung piles; adults emerge up to 8 weeks later. Females are active from mid-July through August.

The fourth species of economically important bot flies is the sheep bot fly, *Oestrus ovis* (Linnaeus). Other common names include the sheep nose bot and the sheep gadfly. Peak female fly activity is in June and July, although they are present from May to late August. Females are most active during the heat of the day, depositing newly hatched white larvae singly in the nostrils of their hosts (eggs hatch within the female). The larvae move through the nasal passages feeding on mucous secretions until they reach the sinuses where they attach to the walls. The larvae mature within 10 months, then crawl out the nasal passages, drop to the ground and pupate in the soil. After about 6 weeks, adults emerge, mate, and females quickly commence depositing larvae.

Sheep bot fly larvae in nasal sinus of host

Gordon Surgeoner
University of Guelph

Food Hosts and Damage

Horse, nose and throat bot flies attack horses, mules and donkeys, with the horse bot the most common of the three species. Female nose and throat bot flies are the most annoying to horses because of their egg-laying behavior. Horses can lose condition as a result of interrupted grazing or suffer physical injury trying to evade the "strikes" of the persistent females. Young larvae can cause irritation and pus pockets when burrowing into the mouth tissue. The presence of large numbers of larvae in the digestive system can cause loss of condition, reduced weight gains and vitality.

Sheep bot flies attack sheep and goats. Bunching, head shaking, feet stamping and rubbing noses against the ground are typical behavioral responses of sheep and goats to the presence of female bot flies. Normal grazing patterns are interrupted which can lead to reduced weight gains. The activity of young larvae can damage the nasal membranes, causing excessive nasal discharge (commonly called "snotty nose") and occasional bleeding. Larvae in the sinuses cause symptomatic frequent sneezing and distressed breathing. The eyes may become inflamed and the animal may step higher and stagger about with the head held high. Frequent sneezing is observed when the mature grubs are moving down the nasal passages prior to pupation.

Control

Bot flies attacking horses can be controlled either at the egg or larval stage. Washing the eggs of the horse bot fly with warm water will induce hatching and the larvae can be washed away. The most effective method is to treat horses orally with a larvicidal compound that kills larvae in the digestive tract.

Sheep can be protected from sheep bot flies by providing darkened shelters with curtained doorways into which the sheep can escape the flies. Applications of greese or pine tar to the sheeps' noses during the fly season prevents larvae from being deposited. An oral drench is registered for the control of bot larvae in the sinuses.

Common Cattle Grub
Hypoderma lineatum (de Villiers)

Northern Cattle Grub
Hypoderma bovis (Linnaeus)

Appearance and Life History

Cattle grubs are the larvae of two species of warble flies found across the prairies, and are among the most injurious of cattle pests. Adult flies, which somewhat resemble honey bees in size and appearance, are also called gad flies, heel flies or bomb flies because of their egg-laying behavior. The white eggs are elongate and measure about 2 mm in length. Mature larvae or grubs are dark grey and measure 25 mm. Puparia containing the pupae are also dark grey in color and somewhat smaller than mature larvae.

Northern cattle grub larva and breathing hole

M. Herbut

Adults of the common cattle grub emerge from puparia in May and June; mating and egg laying occur soon after. Females of this species do not excite their hosts during egg laying as much as female northern cattle grubs, and are able to lay their eggs in rows of 3-10 along a single hair. Each female may deposit 400-800 eggs during her lifetime. Eggs hatch in 2-7 days, and the young larvae crawl down the hairs and enter the skin, usually through a hair follicle. After 1-2 months of migration in the host's body, they appear in the submucosa of the esophagus resembling small grains of rice. From the esophagus the grubs migrate through the host's body and appear along the back about 7-9 months later (late December to late March). Upon reaching the back, the grubs cut breathing holes in the hide and form a small pouch called a warble in which they feed upon tissue secretions for 4-10 weeks. Common cattle grubs begin to drop out of the warbles during the second week of March, and most will leave the warbles by late June. Upon dropping to the ground, the grubs pupate amongst the soil debris, and adults emerge 1-3 months later.

The northern cattle grub has a similar life cycle. Adults begin emerging from puparia from mid- to late April. Females annoy cattle by their incessant buzzing about the animals and darting onto the animals to lay their eggs singly on hairs. Each female lays 400-800 eggs during her lifetime. Eggs hatch in 2-7 days and the young larvae enter the body of their

host similar to common cattle grubs. Northern cattle grubs migrate through the body for 1-2 months before appearing in the tissue surrounding the spinal column. From late January to early May, the grubs reach the back where they cut breathing holes and form warbles. The grubs drop from the back 4-5 weeks later than the common cattle grubs, with most having dropped by the end of June.

Adults of both species are not known to feed but live on the energy stored in the earlier stages of development. The first appearance of grubs is generally 4-6 weeks earlier on the southern prairies than in the northern regions.

Food Hosts and Damage

Cattle are the preferred hosts of both cattle grub species, however, horses, buffaloes, goats and even humans have been infested. In horses, the grubs appear in the saddle area, and hence can cause severe irritation if the infested horses are saddled for work or pleasure.

All stages in the life cycle of cattle grubs except the pupa do cause production losses. Adults, especially northern cattle grubs, cause cattle to gad and seek shelter in shade and water instead of grazing. Gadding causes reduced weight gains and loss in milk production, induces abortions, leads to exhaustion and possible injury, and causes inefficient use of pasture at a time that pastures are quite often most productive. Young grubs entering the hide cause rashes and sores that become infected easily. Some injury or discomfort occurs when the grubs appear along the esophagus and spinal cord. As grubs make respiratory and exit holes in the back, cattle weight gains decrease and the damaged hides are discounted at slaughter. When the grubs are in the back, the warbles and surrounding damaged tissue must be trimmed off during processing, resulting in downgrading of the carcasses.

Control

At present there are no effective mechanical or cultural procedures to control cattle grubs in pastured or range

cattle. Grubs can be squeezed out of the warbles by hand or killed with contact or systemic insecticides applied to the hosts. Systemic and injectable insecticides are the most effective and practical means of controlling grubs in beef and non-lactating dairy cattle. Systemics should be applied after adult fly activity ceases in late summer, and should not be applied during December, January or February to avoid host-parasite reactions due to the death of grubs in the esophagus and spinal column.

The prevalence and severity of cattle grub infestations have decreased dramatically across the prairies with the widespread use of systemics and the promotion of cattle grub control.

Face Fly
Musca autumnalis DeGeer

Appearance and Life History

The face fly is an introduced non-biting pest of cattle found in most cattle-producing regions. Female face flies closely resemble house flies in size and appearance; males have a median dark band down the abdomen and are orange-yellow along the sides. Eggs are white and measure 0.5 mm wide by 3 mm long, and have a dark respiratory mast or projection extending about 0.7 mm from one end. The larvae resemble house fly maggots; puparia are white, measuring 6-10 mm in length.

Face flies on face of Hereford steer

Gordon Surgeoner
University of Guelph

Three to 6 days after emerging from puparia, female face flies mate and then seek sources of protein required for egg development. Both sexes feed on carbohydrates from plants and dung throughout their adult life. Each female lays an average of 20 eggs in fresh cattle dung every 2-8 days during about 21 days, for an average of 100 eggs per female life span. Eggs hatch in 18-24 hours, and the larvae mature in 3-5 days. Eggs are not laid in dung where cattle are confined in pens or drylots, and are never found in dung inside buildings. Mature larvae migrate away from the dung to pupate; adults emerge 7-10 days later. Under pasture conditions, the life cycle requires 17-21 days to complete (the lower the temperatures, the longer to complete the life cycle). Up to four generations can be produced from early spring to early fall. As fall approaches, young sexually immature flies leave the pastures and aggregate in overwintering sites such as hollow trees, church steeples, barns, houses and other buildings. Flies will overwinter in the same buildings year after year. The overwintered flies emerge the following spring and return to the pastures in search of hosts. The adults are strong fliers and can travel several kilometres in a day.

Food Hosts and Damage

Face flies feed on the tears, nasal mucus and saliva secreted by cattle. These flies will also feed to some extent on the same secretions on horses. They will also feed on blood, amniotic fluids, afterbirth, and milk on calves' faces. The majority of face flies (70-95 percent) found on cattle are females as they have a greater requirement for protein than males. Also, only 5 percent of the face fly population in a pasture is on the cattle at any one time. They are most active on bright, warm days with little wind.

The feeding activity of face flies can damage cattle eye tissue, and face flies can transmit *Moraxella bovis*, the organism responsible for pinkeye. In residences harboring overwintering face flies, the flies may become pests during the winter if they get into rooms during warm days.

Control

There are no cultural control methods to protect cattle from face flies. Several insecticides are registered for application to cattle for protection against this pest using various application techniques. Because only a small proportion of the population in a pasture is present on cattle at any one time, cattle should have access to continuous treatment (backrubbers, insecticidal ear tags) to effectively reduce the numbers of flies on the cattle.

Horn Fly
Haematobia irritans (Linnaeus)

Appearance and Life History

The horn fly is a common and widespread introduced pest of range cattle. Adult flies are dark grey and about half the size of house flies. Eggs are reddish-brown and about 1 mm long. Larvae resemble house fly larvae, and measure up to 8 mm when mature. Pupation occurs inside puparia which resemble house fly puparia but only measure 3-4 mm long.

Horn flies resting on back and withers of Hereford bull

M. Herbut

Adult horn flies emerge from overwintered puparia in late May and June and immediately seek cattle hosts upon which to feed and rest. After mating and taking a blood meal, each female fly will deposit up to 400 eggs in batches of 20-24 under fresh cattle dung during her life span of 3-7 weeks. The eggs hatch in 1-2 days, and the larvae feed within the dung for 4-10 days, depending on the temperature. Pupation occurs within the dung or in the soil below it, and adults

emerge from the puparia in 5-13 days. The egg-to-egg life cycle of the horn fly is 3-5 weeks, again depending on temperature. Up to four generations can be produced from June to the first killing frost in the fall; peak numbers occur in July and August. With the onset of cold weather, mature larvae and pupae overwinter in or under the dung and adults emerge the following spring.

Food Hosts and Damage

Both male and female horn flies bite cattle and feed on blood. They remain on the hosts day and night, resting head downwards on the shoulders and backs of cattle, and feeding along the undersides of their hosts by piercing the hide and sucking the blood. Female flies will leave the host to lay eggs under freshly dropped dung. Feeding damage appears as crusty, scabby sores along the underline of hosts. The flies feed around the edges of the previous wounds, causing the sores to get larger each successive season of attack. Horn flies also transmit small filarial worms that infect the sores, however, the consequence of their presence in the sores is not well known.

Horn flies feeding along underline of host

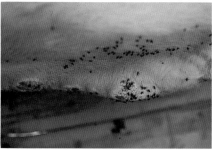

H. Philip

Each fly may feed 10-38 times a day, causing irritation and annoyance to the hosts. Research has shown that infested yearlings on irrigated pasture can lose 18-45 percent of their potential weight gain during the horn fly season. Other losses reported include blood loss up to 7g/head/day, milk loss up to 50 percent of daily production, and reduced weaning weights of calves. Horn flies have also been associated with pinkeye disease in range cattle.

Horn flies have shown a preference for dark-colored cattle and usually stay on black areas of black and white cattle. The flies will move from animal to animal within a herd, and rarely leave the herd. Bulls are more attractive to horn flies than cows or yearlings. Up to 4000 flies can be found on a bull in mid-summer, whereas cows and yearlings will carry only up to 600 flies per animal. Calves under 4-5 months of age are not attacked. Horn flies do not occur in damaging numbers on cattle maintained in feedlots or other confinement facilities.

Control

Complete and continuous protection of range cattle is required to prevent losses due to horn fly attack. Insecticides provide the only effective means of protecting cattle, and should be applied at the first appearance of the flies in the spring. Insecticides can be applied as powders, whole-body sprays, forced-use and free-choice self-treatment backrubber and insecticide-impregnated ear tags. Ear tags must be applied as per label instructions; free-choice backrubbers must be located where cattle congregate and be properly maintained; forced-use backrubbers can be hung across gates into fenced-off water dugouts or salt licks. By controlling the flies emerging in the spring, a damaging population will not develop in the herd and production losses will be kept to a minimum.

Horse Flies
Tabanus spp., *Hybomitra* spp.

Deer Flies
Chrysops spp.

Appearance and Life History

Horse and deer flies are common on the prairies as biting pests of livestock and humans. Adult horse flies are medium to large robust flies, measuring 5-25 mm in length, brown to black in color, sometimes with yellowish or orange stripes or bands on the abdomen. The wings are transparent. Adult deer flies are smaller (about 6-10 mm long), more delicate flies, lighter colored, often with

dark patterns on the transparent wings. Both horse flies and deer flies have brilliantly colored eyes which are banded, spotted or striped with rainbow colors. Eggs are elongate and dirty white to black in color. The larvae are elongate in form, tapering sharply at both ends, with marked ridging at each segment, and measure 10-30 mm when mature, depending on species. The puparia are brown in color.

Horse flies feeding on Hereford cow

Department of Entomology
University of Manitoba, Winnipeg

Adult horse and deer flies appear in June and July, with some species present in August. After mating and taking a blood meal, females lay several hundred eggs encased in dirty white to dark-colored waterproof cases attached to vegetation and other objects overhanging or in water and swampy areas. About a week later, the very small larvae or maggots hatch and fall into the water or soft mud along the margins of streams, ponds or swamps. Larval development time varies according to environmental conditions and species. The larvae overwinter in their selected habitats and mature the following spring. Pupation occurs in drier areas and lasts 1-3 weeks before adults emerge. Most horse and deer fly species have one generation per year. Adults are most active on hot, sunny days (they are not active at night) and are most abundant near breeding sites. They are strong fliers and will seek hosts some distance from their origin.

Food Hosts and Damage

Horses and cattle are the preferred hosts of horse and deer flies, however, other mammals including humans can be attacked. Blood-seeking female flies are attracted to moving objects and dark shapes, hence dark-colored animals usually face greater attacks. Male flies feed only on nectar, honeydew and other liquids. Female mouthparts are stout and blade-like, and inflict deep, painful wounds from which they sponge up the escaping blood for several minutes. Some blood may continue to escape after the female has left which is fed upon by other fly species. Horses and cattle pastured near horse and deer fly breeding sites can suffer significant losses in condition from annoyance and blood loss, especially during June and July. Affected animals will seek deep-shaded areas in trees or buildings to escape fly attack, and injury may result if the animals stampede. Horse flies are known vectors of anthrax and anaplasmosis, and deer flies are important vectors in the transmission of tularemia among rodents and several fur-bearing animals.

Control

Horse and deer flies are very difficult to control by chemical and cultural means. Commonly available ready-to-use sprays lack sufficient "knock-down" power to prevent the flies from biting and feeding. Backrubbers used for horn fly control may provide some protection and relief to cattle. Horses should be stabled or allowed access to deep shade during days of intense fly attack. Drainage of breeding sites is not a practical nor environmentally acceptable control procedure.

House Fly
Musca domestica (Linnaeus)

Appearance and Life History

The common house fly is one of the most familiar insect pests of humans. Adult flies measure 5-12 mm in length, with four dark stripes down the thorax, yellowish spots on each side of the abdomen, and transparent wings. Eggs are slightly curved, about 1 mm long and white in color. Larvae, or maggots, are whitish, tapered from hind end to front, and measure up to 15 mm long when mature. The pupal stage is passed in a puparium, a reddish-brown, smooth, barrel-shaped structure about half the size of a wheat kernel.

Within 2 days of emergence from puparia, female flies mate and egg development commences. Each female can lay batches of about 120 eggs every 2-4 days for 3 or more weeks, depending on weather conditions and food supply.

Adult house fly

M. Herbut

Eggs are laid in cracks and crevices in moist, decaying organic matter upon which the larvae will feed. The eggs hatch in 1-2 days, and the larvae feed for 3-8 days before seeking a drier, undisturbed site to pupate. The pupal stage lasts 3-10 days, and the entire life cycle can be completed in 10-14 days under warm conditions (25-30°C). Outdoors, up to 10 generations can be completed from May through September. House flies overwinter as larvae or pupae in breeding media that do not freeze because of domestic heating or microbial fermentation. Overwintering by adult flies is not considered significant. House flies are not migratory, but generally remain within 500 metres of the breeding site. However, flies will move up to 5 km in search of breeding sites if overcrowded or short of food.

Food Hosts and Damage

Because of their feeding and breeding habits, house flies have the potential to transmit human and livestock disease organisms. Besides their public and animal health importance, they are also an annoyance to people, especially to people living near poorly managed livestock or poultry facilities. House flies are not a significant annoyance to livestock. Because of their habit of defecating on surfaces while or soon after feeding, they can quickly foul the interior of livestock confinement facilities, and contaminate food handling and preparation equipment in homes and restaurants.

Control

House fly population development is limited by such factors as weather, predators, parasites, diseases, adequate food and breeding sites. The use of biological agents such as parasitic wasps to control house flies in livestock and poultry confinement facilities has had limited success in research studies in Canada. On the prairies, the key to successful fly control is the elimination or reduction of breeding media, or preventing fly access to the breeding media. The use of insecticides will not reduce fly numbers to acceptable levels if proper waste management practices are not used at the same time. Constant reliance on insecticides can result in the house fly population developing insecticide tolerance or resistance, especially if only one product is used repeatedly. A selection of two or three insecticides of different classes (carbamates, organophosphates, synthetic pyrethroids) should be used in rotation to prevent resistance development. Insecticides can be applied as sprays to breeding media such as manure piles, and to surfaces where flies congregate. Insecticide baits can be applied as scatter baits along alleyways and on window sills, or can be applied as ready-to-use bait cards or bands that are tacked to walls or ceilings where flies congregate. Electronic traps containing ultraviolet lights are useful indoors to capture flies which enter buildings. Such traps are less effective if they are located in buildings containing breeding media. Screens placed over windows and air vents, and screen doors, will reduce fly entry into buildings.

Mosquitoes
various species

Appearance and Life History

There are four genera of mosquitoes which attack warm-blooded animals on the prairies – *Aedes, Anopheles, Culex* and *Culiseta*. Eggs are elongate, about 0.6 mm long in most species, and are dark brown or black when mature and ready to hatch. Larvae or "wrigglers" are elongate, soft-bodied, dark-colored, 5-10 mm long, with a head, swollen thorax and a cylindrical abdomen. Small tufts of bristles occur at various points along the abdomen. Larvae of many species have a breathing tube or "siphon" at the end of the abdomen through which to breath at the water surface. The pupae or "tumblers" are comma-shaped. The rather large dot of the comma is made up of the head and thorax while the tail is formed by the flexible abdomen. Pupae breath by means of a pair of projections or "trumpets" on the back of the thorax.

The life history of the four genera are very similar except for the means of overwintering and egg-laying behavior. *Aedes* species overwinter as eggs deposited around the margins of sloughs, above the water line in ditches, or in depressions in grassy terrain that are subject to flooding due to snowmelt or rainfall. *Anopheles, Culex* and *Culiseta* species overwinter as mated females (the males do not overwinter), and eggs are laid either singly (*Anopheles*) or in a raft (*Culex* and *Culiseta*) on the surface of permanent or semi-permanent bodies of water. After eggs hatch, the aquatic life history is the same for all genera.

Aedes species are divided into two groups – the spring group and the summer group. Eggs of the spring group hatch in the cold water of snowmelt pools, and there is usually one generation per year. Adults emerge during the spring and early summer, and after mating and taking a blood meal, females will lay batches of up to 250 eggs. The blood meal and egg-laying cycle may be repeated three or more times before the death of the female. Eggs laid by spring *Aedes* will not hatch until the following spring. Eggs of the summer group only hatch when immersed in warm water and there can be up to four generations each summer. Therefore adults of this group appear later than those of the spring group and can be a problem for the remainder of the summer. Blood-fed summer *Aedes* females lay their complement of eggs above the waterline of temporary water bodies that are drying up. The blood meal and egg-laying cycle can be repeated several times. Following the 4-6 days required to complete embryonic development, eggs can hatch within 5 minutes of immersion in pools created by rainfalls over 2.5 cm, runoff, or irrigation. The larvae can mature, and pupate, and adults emerge within 10 days under very warm conditions and ample food supply. Thus under warm, wet conditions, summer mosquito populations can build up quickly. Under dry conditions, eggs can remain viable for 2-3 years. Summer *Aedes* adults live for only a few weeks unlike the spring adults which can live 2-3 months. Towards the end of summer and into early fall, shorter day length and lower temperature cause the females to lay eggs that, like the spring species, require a freezing period before they will hatch.

Female *Anopheles*, *Culex* and *Culiseta* overwinter in protected sites such as hollow logs, rodent burrows, rock piles and buildings. Several generations may be produced during the summer. Females that emerge in late summer mate, but do not seek a blood meal before seeking overwintering sites.

Larvae feed by filtering small organic particles from the water (microorganisms, algae, pollen, fungal spores) and by nibbling on algae, decaying plant matter and other suitable material. Larvae are active swimmers, and while they generally remain near the water surface, they will "wriggle" or swim down to the bottom when disturbed. Pupae do not feed, but remain at the water surface to breathe until disturbed; then they will "tumble" away.

In general, mosquito larvae are rarely found in large bodies of water which are subject to wave action. They are never found in fast-moving water but some may be found in backwater or in very slow-moving canals or streams. Mosquitoes generally prefer stagnant water as breeding sites. In general, female mosquitoes do not disperse far, tending to remain within 5 km from their emergence site (disperal up to 25 km is possible however). High winds and temperatures discourage adult activitiy, and most tend to be more active during the evening and morning hours when winds and temperature are usually lower.

Mosquito breeding site in pasture

Alberta Government Photo

Food Hosts and Damage

Adults of both sexes feed on flower nectar and various juices of fruits but the females also take blood meals to provide protein for egg development. Feeding is accomplished by using piercing and sucking mouthparts. All warm-blooded animals are subject to attack by blood-seeking female mosquitoes that are attracted to their hosts by the carbon dioxide given off through respiration as well as by visual stimulii. *Aedes* species are usually the most numerous and annoying to humans. *Culiseta* and *Anopheles* generally prefer large animals such as moose, deer and livestock. *Culex* and some *Anopheles* attack birds as preferred hosts. Mosquitoes can be a limiting factor in the outdoor recreational and work activity of people, especially during summers of above-normal rainfall. Livestock can be seriously stressed during periods of intense mosquito activity, causing reduced weight gains and milk production due to interrupted grazing, and, in extreme cases, physical injury may occur as a result of animals stampeding to escape attack. Cattle and horses will characteristically bunch up or seek non-grassed areas in pastures to avoid attack. *Culex tarsalis* transmits sleeping sickness (western equine encephalomyelitis) to horses and humans; the disease can be fatal to both hosts. Mosquitoes will also invade homes and livestock and poultry confinement facilities in search of blood meals.

Control

Various cultural and chemical methods for controlling mosquitoes or reducing mosquito attack are available. Cultural controls include elimination of mosquito breeding sites through drainage of water from depressions in pastures, from roadside ditches and other sites of temporary water collection. Where irrigation is used for crop production, water should not be allowed to accumulate in small pools; canals should be kept free of debris and vegetation and should be drained completely when not in use. Canal seepage should be prevented to eliminate escape of water that could pond or accumulate in ditches. In pastures, keeping a small area free of vegetation provides an area for livestock to congregate and find some relief from mosquito attack. Screened doors, windows and ventilation ports will prevent mosquito entry into buildings.

Several chemicals are registered for application to mosquito breeding sites to kill the larvae. Chemicals can be applied to cattle and horses for protection against mosquitoes, and repellants are available for human protection. Adult mosquitoes can be controlled by treating infested areas with residual sprays or by fogging. Keeping lawns mowed and eliminating unwanted vegetation around buildings and shelterbelts will aid in reducing mosquito annoyance. In some areas, farmers have formed mosquito control associations to mount a cooperative attack against mosquitoes and provide area-wide control. Control efforts by individual producers often fail because of the immigration of mosquitoes from adjacent uncontrolled areas.

Rat-tailed Maggot

Eristalis tenax (Linnaeus)

Appearance and Life History

The rat-tailed maggot adult, commonly known as a drone fly, is a large, robust fly that resembles a honey bee in size and color. Larvae, called rat-tailed maggots because of the long respiratory siphon or tube extending from the hind end, are legless, whitish, soft-bodied and measure 20 mm long when mature (excluding the "tail"). Puparia are hard-bodied, grey in color, about 15 mm long, and retain the breathing tube.

Drone fly larva or rat-tailed maggot

Alberta Government Photo

Female drone flies lay their eggs in the spring on the surface of stagnant water rich in organic matter such as manure settling ponds or tanks, sewage ditches, latrines and stagnant livestock water troughs. Larvae hatch within a week and commence development. No information is available on the development time of the life stages. When mature larvae are ready to pupate, usually in late July and August, they leave the stagnant water to pupate in drier and protected situations such as behind walls and under debris and equipment on floors. It is suspected that adult flies emerge after 3 weeks and immediately seek overwintering sites. It is also possible that this insect overwinters in the pupal stage and adults emerge the following spring. Within 1 week of emergence, female flies commence egg laying. Only one generation is produced each year.

Food Hosts and Damage

Rat-tailed maggots are only a pest because of their presence in livestock confinement facilites, especially at the time they are migrating in search of pupation sites. Maggots feed on organic matter in the stagnant water; adults feed on the nectar of flowers over which they are often seen hovering. Adults are never numerous enough in or around buildings to be a nuisance.

Control

Since rat-tailed maggots are of no known economic importance, there is no registered chemical control. Infestations can be reduced by preventing egg-laying females access to manure settling tanks and other breeding sites in buildings. Regular flushing and cleaning of manure collection and storage facilities every 2-3 weeks, especially in the early summer, will remove any eggs and maggots present. Collection and destruction of the maggots when they are found on the floors will reduce the number of subsequent adults and thereby the intensity of the infestation the following year.

Sheep Ked

Melophagus ovinus (Linnaeus)

Appearance and Life History

The sheep ked is a wingless blood-sucking fly that attacks sheep. Adult keds are often referred to as sheep "ticks" because of their flat, reddish-brown color and tick-like shape. They measure 5-8 mm long. Larvae appear as a whitish object, about 2-3 mm long, within the abdomen of female keds. Puparia are shiny, hard, chestnut-colored and somewhat resemble radish seeds in size and shape.

Adult sheep ked

M. Herbut

Adult female keds start to reproduce 14-30 days after emergence from the puparia. Each female produces one offspring every 7 or 8 days for a total of 12-15 offspring over an average lifetime of 4 months. After mating, a single fertilized egg hatches within a uterine pouch in the female's abdomen. The larva feeds on uterine secretions, and just prior to pupation, is extruded from the abdomen. The female secretes a glue-like substance over the larva which fastens the larva onto the wool. Pupation occurs almost immediately, and the puparium remains attached to the wool until the adult emerges 18-40 days later. Adults and pupae separated from hosts will die in 4-7 days. Reproduction is continuous, and several generations can be completed in a year. Populations are generally low during the summer and gradually increase with the onset of lower temperatures in the fall.

Food Hosts and Damage

Sheep are the primary hosts of sheep keds, but goats will occasionally be attacked. Both sexes feed by piercing the skin and sucking blood, causing the sheep to rub, bite and scratch at the wool to relieve the irritation. This defensive behavior can damage the wool coat. A small number of keds does not stress the animals; large numbers can reduce the vitality and growth rate of infested animals. Wool can be soiled and discolored by ked excrement resulting in lower sale prices.

Control

Shearing will remove some adults and puparia but enough are left unharmed to maintain and increase the population. The only effective means of ked control is through insecticidal treatment.

Stable Fly

Stomoxys calcitrans (Linnaeus)

Appearance and Life History

The stable fly is a blood-feeding fly commonly found around livestock confinement facilities. Adults closely resemble house flies, but are easily distinguished by the prominent piercing and sucking mouthparts that protrude from the front of the head. They are brown to grey in color and in bright sunlight the body has a greenish-yellow sheen. The abdomen has a checkered appearance. Eggs are about 1 mm long. Mature larvae are creamy white, about 10 mm long, and resemble house fly larvae. Puparia are 6-7 mm long and chestnut-brown in color.

Adult stable fly

M. Herbut

Female stable flies mate about 2-3 days after emergence from the puparia and immediately seek blood meals for egg development. Egg laying occurs about 9 days after emergence and each female lays up to 90 eggs each egg-laying cycle. One female may lay more than 1000 eggs in her 3- to 4-week lifetime. Favored breeding media include manure mixed with soil, straw, hay silage or grain, rotting vegetation such as old hay and straw stacks, moist silage, decaying lawn clippings, and decaying plant material along lake shores. Depending on moisture, temperature and availability of food, larvae mature in 10-21 days. Pupation occurs in drier areas on the breeding media, and adults emerge 6-26 days later depending on temperature. The entire life cycle can be completed in 3-4 weeks. Stable flies overwinter as larvae and pupae in breeding media kept from freezing by microbial fermentation or domestic heating. Several generations can be produced each summer, with numbers increasing as the summer progresses.

Food Hosts and Damage

Stable flies feed on a wide range of domestic and wild animals, but are of economic importance for their attacks on livestock and humans. Both sexes take blood meals by piercing the skin or hide and sucking up the blood. Animals

maintained in or near confinement facilities are most susceptible to attack during the daylight hours. Large numbers of stable flies are a great annoyance and irritation to cattle, and significant losses in milk production and weight gains can result. Preferred sites of attack on animals include the legs, belly and sides. The flies will enter barns and dwellings, and attack animals or humans, inflicting painful bites. Stable flies are strong fliers and will disperse widely seeking hosts and breeding sites.

Control

Elimination of breeding media provides the most effective outdoor control of stable flies around farmsteads, residences and beaches. Breeding media should be burned or spread thinly to dry. Indoors, space sprays or aerosols can be used. Insecticides are available for direct application to animals, and repellants are effective for short periods to protect humans. Limiting fly access to barns or stables in combination with residual insecticide wall sprays will reduce annoyance to confined or stabled animals.

Heteroptera
Bed Bug
Cimex lectularius Linnaeus

Appearance and Life History

The bed bug is a wingless blood-feeding insect. Adults are oval, flattened, reddish-brown, and measure 4-5 mm long and 3 mm wide. Eggs are elongate, about 1 mm long, pearly white and opaque. Nymphs resemble adults in color and shape but are smaller in size.

Adult bed bug

M. Herbut

After mating, female bed bugs deposit their complement of about 350 eggs at a rate of 2-3 per day over their life span of 8-18 months. Eggs are cemented to surfaces in undisturbed places such as crevices in the furniture or walls of rooms in which people sleep. Nymphs hatch from the eggs in about 10 days and undergo five molts before reaching maturity in 37-128 days, depending on food supply and temperature. Blood meals are required before each molt and for egg production. All stages of development can be found in occupied dwellings as reproduction is continuous year round.

Food Hosts and Damage

Nymphal and adult bed bugs feed on warm-blooded animals, most notably humans, by piercing the skin and sucking the blood using their tube-like mouthparts. Nymphs feed at about 10-day intervals and adults feed weekly, each blood meal taking about 5-10 minutes to complete. The degree of irritation of the bites varies with individuals. A red welt and local inflammation will occur at the site of each bite as a result of the saliva injected by the bed bug while piercing the skin. Bites may become infected if scratched. Nymphs can survive starvation for up to 4 months and adults up to 8 months at room temperature (18-20°C).

Bed bugs hide during the day in cracks and crevices in bedrooms, emerging shortly before dawn to feed on unsuspecting victims. After feeding, they will defacate dark-colored faeces on surfaces near their hiding places. When disturbed, they will emit a characteristic odor due to the secretion of so-called "stink glands". Bed bugs are not restricted in their distribution to older, unkept, occupied dwellings. Infestation of new houses is almost exclusively due to bugs being carried on people's possessions such as on outer clothing, luggage and laundry, or they may be introduced with used beds, bedding and furniture. Bed bugs will not move far from bedrooms or sleeping quarters unless vacated, and even when unoccupied, the bugs will remain for long periods awaiting new hosts. In multiple-family dwellings, bed bugs can move between rooms, especially in ill-constructed buildings.

Two other species of blood-sucking bugs have been recorded biting humans – the bat bug, *Cimex pilosellus* (Horvath), and the swallow bug, *Oeciacus vicarius* Horvath. Bats are the usual host for bat bugs and swallows for swallow bugs. Both species will invade dwellings in search of alternate hosts when their natural hosts vacate their roosts or nests. Both can be distinguished from the common bed bug by the presence of long body hairs absent in bed bugs.

Control

The presence of bites and of faecal deposits on walls and bedding are evidence of bed bug infestations. Rooms in which bites have been inflicted should be thoroughly searched during the day for bed bug hiding places. Spraying cracks and crevices with a household aerosol insecticide will irritate the bugs and drive them out of the hiding places. Infested clothing, furniture and bedding should be thoroughly cleaned to remove eggs and feeding stages. Chemicals are registered for control of this pest. Preventing bats from roosting in or swallows from nesting on dwellings will eliminate the risk of bat or swallow bug infestations.

Hymenoptera
Yellowjackets, Paper Wasps, and Hornets

Appearance and Life History

Yellowjackets, paper wasps, and hornets are beneficial insects that attack and destroy harmful insects. Several species of these social insects are found on the prairies. Adults may be black with yellow markings or black with white markings. Wings are translucent, dark and folded lengthwise when the insects are at rest. Adults range in length from 12 to 25 mm, depending on the caste or type. A colony consists of three types: queens (the largest individuals), workers (sterile females that make up the largest numbers), and drones (males). Eggs are milky white, 1-2 mm long, and sausage-shaped.

Larvae are creamy white legless grubs that range in size from slightly larger than the egg at hatching to adult size at maturity. Larvae are usually found only in the nest cells. Pupae resemble adults in size and shape but are whiter in color.

Yellowjacket wasp worker

M. Herbut

Wasps and hornets overwinter as mature, fertilized queens in protected locations such as under bark, in stumps or in hollow logs. In the spring, the queen emerges from hibernation, finds a suitable nest site, and begins building a nest by chewing wood fibers into a pulpy mass. She constructs a small pedicel, a canopy and a few hexagonal cells. She deposits a single egg in each cell. The queen continues to forage and construct more cells and enlarges the canopy until the eggs begin to hatch in about 2 weeks. The newly hatched larvae remain in the cell and the queen forages for food for the young and herself, and for more building material.

Larvae feed for 2-3 weeks before pupating in the cells. The first wasps to emerge are small, sterile female workers who take over nest construction, foraging and brood care from the queen. The queen eventually restricts her activity to egg laying. Colony expansion then occurs very rapidly until several layers of comb are enclosed in a paper nest. Depending upon the species of wasp, availability of food, and weather conditions, maximum nest diameter may reach 40 cm with as many as 5000 insects. In August, reproductive cells, which are larger than the worker cells, are built and new males and queens are produced. These emerge in the early fall and mate. The males soon die but the queens seek out hibernation sites to overwinter. The abandoned nests disintegrate during the winter and are never used again.

Nests may be built below the soil in mouse burrows, between the walls of houses or as aerial nests in trees, sheds, or under eaves of houses. Nests built of wood fiber resemble paper and are completely enclosed except for a small opening at the bottom. Several horizontal combs of cells are suspended one below the other inside the nest. Years with warm dry springs usually have more wasp problems in August than years with cool wet springs.

Food Hosts and Damage

Most wasps and yellowjackets are hunters that feed on caterpillars, house flies and other insects. They are especially useful in controlling many ornamental pests. A few species, however, scavenge for meat and liquid sweets and can reach pest status that may range from mild nuisance to severe hazard. Barbecued meat, jams and soda pop are very attractive food material. Wasps are capable of inflicting a painful sting and are a serious nuisance around homes and recreational areas. They are capable of stinging repeatedly when annoyed and may attack in force if their nest is endangered. The stinger is a needle-like projection at the end of the abdomen that injects a venom which causes mild to very severe pain in the area of the sting. People who have high sensitivity to the stings should avoid areas where these pests are found and seek immediate medical attention if stung.

Control

Controls are not necessary unless wasps are a nuisance or a threat to human health. Hanging a piece of meat over a bucket of soapy water may help reduce wasps in an area. Wasps trying to fly away with bits of meat will fall into the water and drown. Cooked meat saturated with an insecticide and hung from a tree is also effective as the poisoned meat is carried to the nest where it kills all members. Poisoned-bait stations with ground meat and powdered insecticide may be effective in controlling wasps when the number and location of the nests are not known.

Aerial nests may be sprayed with special aerosol products that propel the insecticide up to 3 metres. Other insecticides directed at the opening are also effective. Pour insecticides into ground nests and plug the entrance with insecticide-soaked cotton or steel wool. Applications are best done at night when most of the wasps are in the nest and are less active. The person carrying out a control program should be protected with adequate bee veils, clothing and gloves.

Avoidance of wasp stings while working or being in areas where wasps are a problem is of utmost concern to workers in forest and recreation areas. Perfumes, hair sprays, suntan lotions and shaving creams are attractive to wasps and should not be used. Light clothing, the whites and tans, are less attractive to wasps than dark clothing.

Awareness of nest locations is necessary while working outdoors. Calm, slow and deliberate movements in order not to excite the wasps will reduce further stings or threat of stings.

Siphonaptera
Fleas
various species

Appearance and Life History

Fleas are familiar pests of mammals and birds on the prairies. Adults are wingless, reddish-brown to almost black, 2-3 mm long, with long legs adapted for jumping. The body is compressed laterally and armed with numerous backward-pointing bristles and spines. Eyes may or may not be present depending on the species. Both sexes have piercing and sucking mouthparts. Eggs are small (0.5 x 0.3 mm) pearly white and oval. Mature larvae measure 5 mm long, are cream-colored, and are more or less cigar-shaped with a definite head, three thoracic legs and ten abdominal segments sparsely covered with hairs. Pupae are initially white but change to brownish just prior to adult emergence.They have the general shape and appearance of adult fleas, but are soft-bodied.

Adult human flea

Alberta Government Photo

After mating and obtaining a blood meal, the female flea will lay 4-8 eggs on the plumage, hair or clothing of the host. Several hundred eggs can be laid during her lifetime. The eggs usually are brushed off the host and fall on the ground in or near the sleeping place of the host. The eggs hatch in 10 days or less and the young, white legless larvae immediately move about in search of suitable food. The larvae pass through at least three instars over a 1- to 4-week larval period before pupating in silken cocoons. These cocoons are often covered with sand, dirt, dust or other small particles which effectively camouflage the pupae. Under favorable conditions, the pupal period may last 1-2 weeks. However, under adverse conditions, the pupal period may be prolonged up to a year. Adults do not emerge immediately from the cocoon, but wait until vibrations indicating the presence of a possible host stimulate them to emerge and become active. Depending on the species and environmental conditions, a life cycle can be completed in 2-3 weeks or several months.

Adult fleas can survive several months without feeding. Larvae require above 50 percent relative humidity to develop normally and will die under dry air conditions.

Food Hosts and Damage

Adult fleas feed on the blood of mammals and birds using their piercing and sucking mouthparts. Many are host specific, or attack only mammals or birds. Larvae feed on organic debris such as hair, skin scales, scabs and on the faeces of adult fleas found in the sleeping area of hosts. The compressed

shape of the adults allows them to move easily through hair, feathers and garments, and the backward-pointing spines and bristles make dislodgement difficult. Humans are usually attacked at night while asleep. Although the individual blood meal volume is small, repeated feedings can result in significant blood loss and stress to the host. The bites are irritating, itchy and can cause allergic reactions in humans or animals. Secondary infections may result from scratching or nipping at the feeding site. Bites appear as small red spots, surrounded with a red halo, but very little swelling.

The cat flea, *Ctenocephalides felis* (Bouché), and the dog flea, *C. canis* (Curtis), readily feed on humans as well as cats, dogs and certain other mammals. The human flea, *Pulex irritans* Linnaeus, is less frequently encountered on humans and more frequently on swine. Rodents are attacked by various species of fleas which will seek alternate hosts such as humans when their primary hosts die or abandon their nests. The western chicken flea, *Ceratophyllus niger* Fox, occurs occasionally on domestic and wild birds and will bite humans.

Adult chicken fleas

M. Herbut

Control

Since humans are also favored hosts of cat and dog fleas, pet cats and dogs should be kept flea-free to avoid flea infestations. If the pets are frequently noticed scratching or nipping their bodies, treat them with a recommended flea control product applied as a dust, shampoo, or as a flea collar. The sleeping area of the pet must also be thoroughly cleaned and disinfected to prevent reinfestation. Control of rodents under dwellings will eliminate the possiblility of rodent fleas invading residences in search of human hosts. Infested bedrooms should be thoroughly cleaned and vacuumed, and if flea bites are still noticed, residual chemical sprays can be applied to cracks in flooring and along baseboards, cold-air ducts, rugs and under their edges, and under furniture and cushions where fleas hide during the day.

Useful References

1. Borror, D.J., D.M. DeLong, and C.A. Triplehorn. 1976. *An Introduction to the Study of Insects.* 4th ed. Holt, Rinehart and Winston, New York, New York. 852 p. ISBN:0-03-088406-3.

2. Bland, R.G. and H.E. Jaques. 1978. *How to Know the Insects.* 3rd ed. Wm. C. Brown Co. Publishers, Dubuque, Iowa. 409 p. ISBN:0-697-04752-0 (paper) ISBN:0-697-04753-9 (cloth).

3. Martineau, R. 1984. *Insects Harmful to Forest Trees.* Multiscience Publ. Ltd. and Can. Government Publishing Centre, Ottawa, Ontario. 261 p. ISBN:0-919868-21-5.

4. Carr, A. 1979. *Color Handbook of Garden Insects.* Rodale Press, Erasmus, Pennsylvania. 240 p. ISBN:0-87857-250-3.

5. Westcott, C. 1973. *The Gardener's Bug Book.* 4th ed. Doubleday & Co. Inc., Garden City, New York, New York. 689 p. ISBN:0-385-01525-9.

6. Mallis, A. 1982. *Handbook of Pest Control.* 6th ed. Franzak & Foster Co., Cleveland, Ohio. 1101 p. ISBN:0-942588-00-2.

7. Truman, L.C., G.W. Bennett, and W.L. Butts. 1976. *Scientific Guide to Pest Control Operations.* 3rd ed. Harvard Publishing Co., Cleveland, Ohio. 276 p. Library of Congress Cat. Card No. 67-16201.

8. Ebeling, W. 1975. *Urban Entomology.* Univ. of California, Div. of Agric. Sciences, Los Angeles, California. 695 p.

9. Sinha, R.M. and F.L. Watters. 1985. *Insect Pests of Flour Mills, Grain Elevators, and Feed Mills and Their Control.* Publ. 1776 Agriculture Canada, Can. Government Publishing Centre, Ottawa, Ontario. 290 p. ISBN:0-660-11748-7.

10. Williams, R.E., R.D. Hall, A.B. Broce, and P.J. Scholl, editors. 1985. *Livestock Entomology.* John Wiley & Sons, New York, New York. 335 p. ISBN:0-471-81064-9.

11. Ives, W.G.H. and H.R. Wong. 1988. *Tree and Shrub Insects of the Prairie Provinces.* Can. For. Serv., North. For. Cent., Edmonton, Alberta. Inf. Rep. NOR-X-292.

Additional published information on insects is available through local libraries and bookstores. Departments of Agriculture publish and distribute extension leaflets containing the latest insect pest control recommendations. If the publication date of the leaflet is more than 2 years old, check with the Department of Agriculture for any changes to the recommendations.

Index of Common and Scientific Names

Host Index

African Violets
citrus mealybug, 41
cyclamen mite, 47
fungus gnats, 40

Alder
large aspen tortrix, 66
poplar-and-willow borer, 53
twospotted spider mite, 47

Alfalfa
alfalfa looper, 19
alfalfa plant bug, 14
alfalfa weevil, 5
army cutworm, 20
armyworm, 20
beet webworm, 21
blister beetles, 6
lygus bugs, 14
onion thrips, 45
pea aphid, 17
red clover thrips, 46
redbacked cutworm, 26
sugarbeet root aphid, 18
superb plant bug, 15
sweetclover weevil, 10
tarnished plant bug, 14
twostriped grasshopper, 28

Alsike Clover
redbacked cutworm, 26

Apple
boxelder bug, 80
European fruit lecanium, 44
fall cankerworm, 64
forest tent caterpillar, 65
scurfy scale, 44
speckled green fruitworm, 70

Ash
ash bark beetles, 49
ash borer, 61
ash flower gall mite, 48
boxelder bug, 80
carpenterworm, 63
European fruit lecanium, 44
fall cankerworm, 64
forest tent caterpillar, 65
lilac leafminer, 67
scurfy scale, 44

Asparagus
onion thrips, 45

Balsam Fir
eastern spruce budworm, 64

Balsam Poplar
large aspen tortrix, 66
poplar borer, 52

Bananas
Boisduval scale, 44

Barley
army cutworm, 20
armyworm, 20
aster leafhopper, 35
barley thrips, 29, 45
brown wheat mite, 5
clearwinged grasshopper, 27
corn leaf aphid, 16
English grain aphid, 16
grass thrips, 46
greenbug, 16
Hessian fly, 11
June beetles, 50
migratory grasshopper, 28
oat birdcherry aphid, 16
orange wheatblossom midge, 11
pale western cutworm, 25
prairie grain wireworm, 8
Russian wheat aphid, 16
redbacked cutworm, 26
twostriped grasshopper, 28
wheat stem maggot, 13

Basswood
fall cankerworm, 64
linden looper, 68

Beans
European corn borer, 24
June beetles, 50
onion thrips, 45
pea aphid, 17
red turnip beetle, 9
tuber flea beetle, 32

Beet
beet leafminer, 32
imported cabbageworm, 36
sugarbeet root aphid, 18

Begonia
citrus mealybug, 41
green peach aphid, 55

Birch
ambermarked birch leafminer, 57
birch leafminer, 57
bronze birch borer, 49
forest tent caterpillar, 65
late birch leaf edgeminer, 57
linden looper, 68
mourningcloak butterfly, 69
poplar-and-willow borer, 53

Black Spruce
eastern spruce budworm, 64
white pine weevil, 54

Black Cottonwood
poplar-and-willow borer, 53

Blueberry
black vine weevil, 39

Bluegrass
glassy cutworm, 66
sod webworms, 70
wheat stem maggot, 13

Boxelder
boxelder bug, 80
boxelder twig borer, 62
fall cankerworm, 64
linden looper, 68

Broadbeans
blister beetles, 6

Broccoli
cabbage maggot, 33
diamondback moth, 23
imported cabbageworm, 36

Bromegrass
glassy cutworm, 66
wheat stem maggot, 13

Brussels Sprouts
cabbage maggot, 33
diamondback moth, 23
imported cabbageworm, 36
red turnip beetle, 9

Buckwheat
June beetles, 50

Cabbage
armyworm, 20
cabbage maggot, 33

Jack Pine
mountain pine beetle, 51
northern pitch twig moth, 69
white pine weevil, 54

Jade Plant
mealybugs, 41

Juniper
spruce spider mite, 48

Larch
eastern spruce budworm, 64

Leek
onion maggot, 34

Lettuce
alfalfa looper, 19
aster leafhopper, 35
European earwig, 77
greenhouse whitefly, 42
imported cabbageworm, 36
potato aphid, 55
prairie grain wireworm, 8
red turnip beetle, 9
spinach carrion beetle, 31
tuber flea beetle, 32

Lilac
ash borer, 61
fall cankerworm, 64
lilac leafminer, 67
oystershell scale, 44

Lily
gladiolus thrips, 45, 72

Lodgepole Pine
mountain pine beetle, 51
northern pitch twig moth, 69

Manitoba Maple
boxelder bug, 80
boxelder twig borer, 62
fall cankerworm, 64
linden looper, 68

Maple
boxelder bug, 80
carpenterworm, 63
linden looper, 68
scurfy scale, 44

Mayday
forest tent caterpillar, 65

Mountain Ash
ash borer, 61
carpenterworm, 63
pear slug, 58

Mugho Pine
mountain pine beetle, 51
northern pitch twig moth, 69
white pine weevil, 54

Mushroom
fungus gnats, 40

Mustard
beet webworm, 21
crucifer flea beetle, 7
diamondback moth, 23
lygus bugs, 14
pale western cutworm, 25
red turnip beetle, 9
redbacked cutworm, 26
striped flea beetle, 7

Norway Spruce
white pine weevil, 54

Oats
army cutworm, 20
armyworm, 20
aster leafhopper, 35
clearwinged grasshopper, 27
English grain aphid, 16
European corn borer, 24
grass thrips, 46
greenbug, 16
June beetles, 50
migratory grasshopper, 28
oat birdcherry aphid, 16
pale western cutworm, 25
prairie grain wireworm, 8
redbacked cutworm, 26
twostriped grasshopper, 28
wheat stem maggot, 13

Onion
imported cabbageworm, 36
onion maggot, 34
onion thrips, 45
prairie grain wireworm, 8

Orchard Grass
June beetles, 50

Orchid
Boisduval scale, 44

Palm
Boisduval scale, 44

Parsnips
aster leafhopper, 35

Parsley
imported cabbageworm, 36

Peach
boxelder bug, 80
European earwig, 77

Pear
pear slug, 58

Pea
imported cabbageworm, 36
June beetles, 50
pea aphid, 17
spinach carrion beetle, 31

Pepper
onion thrips, 45
tuber flea beetle, 32

Pine
eastern spruce budworm, 64
mountain pine beetle, 51
webspinning sawflies, 59

Plains Cottonwood
poplar borer, 52

Plum
boxelder bug, 80
European fruit lecanium, 44
pear slug, 58

Poinsettias
fungus gnats, 40
greenhouse whitefly, 42

Ponderosa Pine
mountain pine beetle, 51

Poplar
carpenterworm, 63
fall cankerworm, 64
forest tent caterpillar, 65
linden looper, 68
mourningcloak butterfly, 69
poplar budgall mite, 48
poplar gall aphid, 56
poplar-and-willow borer, 53
willow leaf beetle, 50
willow sawfly, 60

Swine
   black flies, 96
   follicle mite, 92
   hog louse, 95
   human flea, 107
   mange mite, 91
   mosquitoes, 102

Swiss Chard
   beet leafminer, 32
   spinach carrion beetle, 31
   sugarbeet root aphid, 18

Timothy
   glassy cutworm, 66
   June beetles, 50
   sod webworms, 70
   strawberry root weevil, 76
   wheat stem maggot, 13

Tomato
   Colorado potato beetle, 31
   European corn borer, 24
   greenhouse whitefly, 42
   imported cabbageworm, 36
   leafminer, 41
   potato aphid, 55
   tuber flea beetle, 32

Trefoil
   pea aphid, 17

Trembling Aspen
   Bruce spanworm, 62
   forest tent caterpillar, 65
   large aspen tortrix, 66
   poplar borer, 52
   speckled green fruitworm, 70

Triticale
   Russian wheat aphid, 16

Turnip
   armyworm, 20
   cabbage maggot, 33
   crucifer flea beetle, 7
   diamondback moth, 23
   millipedes, 79
   onion thrips, 45
   red turnip beetle, 9
   spinach carrion beetle, 31
   striped flea beetle, 7
   turnip maggot, 33

Vetch
   alfalfa weevil, 5
   blister beetles, 6
   pea aphid, 17

Violet
   green peach aphid, 55

Weeping Fig
   brown soft scale, 45

Western wheatgrass
   Hessian fly, 11

Wheat
   army cutworm, 20
   armyworm, 20
   brown wheat mite, 5
   clearwinged grasshopper, 27
   English grain aphid, 16
   grass thrips, 46
   greenbug, 16
   Hessian fly, 11
   June beetles, 50
   migratory grasshopper, 28
   oat birdcherry aphid, 16
   orange wheatblossom midge, 11
   pale western cutworm, 25
   prairie grain wireworm, 8

Russian wheat aphid, 16
   twostriped grasshopper, 28
   wheat midge, 12
   wheat stem maggot, 13
   wheat stem sawfly, 18

White Birch
   bronze birch borer, 49
   European fruit lecanium, 44
   large aspen tortrix, 66

White Pine
   mountain pine beetle, 51

White Spruce
   Cooley spruce gall adelgid, 57
   eastern spruce budworm, 64
   pine needle scale, 43
   spruce spider mite, 48
   white pine weevil, 54
   yellowheaded spruce sawfly, 60

Willow
   Bruce spanworm, 62
   carpenterworm, 63
   cottonwood leaf beetle, 50
   large aspen tortrix, 66
   mourningcloak butterfly, 69
   poplar-and-willow borer, 53
   scurfy scale, 44
   speckled green fruitworm, 70
   willow leaf beetle, 50
   willow redgall sawfly, 59
   willow sawfly, 60

Zinnia
   sunflower moth, 26